Communication Skills in Helping Relationships

A Framework for Facilitating Personal Growth

Communication Skills in Helping Relationships

A Framework for Facilitating Personal Growth

Vonda Olson Long

University of New Mexico

Brooks/Cole Publishing Company

I(T)P™ An International Thomson Publishing Company

Pacific Grove • Albany • Bonn • Boston • Cincinnati • Detroit • London • Madrid
Melbourne • Mexico City • New York • Paris • San Francisco
Singapore • Tokyo • Toronto • Washington

Sponsoring Editor: *Lisa Gebo, Claire Verduin*
Marketing Team: *Nancy Kernal, Romy Fineroff*
Editorial Assistant: *Patsy Vienneau*
Production Editor: *Marjorie Z. Sanders*
Manuscript Editor: *Patterson Lamb*
Permissions Editor: *Cat Collins*
Interior and Cover Design: *Roy Neuhaus*

Cover Photo: *Ed Young*
Art Editor: *Lisa Torri*
Cartoons: *Mark Taylor*
Typesetting: *Scratchgravel Publishing Services*
Cover Printing: *Phoenix Color Corporation, Inc.*
Printing and Binding: *Quebecor/Fairfield*

For more information, contact:

BROOKS/COLE PUBLISHING COMPANY
511 Forest Lodge Road
Pacific Grove, CA 93950
USA

International Thomson Publishing Europe
Berkshire House 168-173
High Holborn
London WC1V 7AA
England

Thomas Nelson Australia
102 Dodds Street
South Melbourne, 3205
Victoria, Australia

Nelson Canada
1120 Birchmount Road
Scarborough, Ontario
Canada M1K 5G4

International Thomson Editores
Campos Eliseos 385, Piso 7
Col. Polanco
11560 México D. F. México

International Thomson Publishing GmbH
Königswinterer Strasse 418
53227 Bonn
Germany

International Thomson Publishing Asia
221 Henderson Road
#05-10 Henderson Building
Singapore 0315

International Thomson Publishing Japan
Hirakawacho Kyowa Building, 3F
2-2-1 Hirakawacho
Chiyoda-ku, Tokyo 102
Japan

Printed in the United States of America

10   9   8   7   6   5   4   3   2   1

Library of Congress Cataloging-in-Publication Data
Long, Vonda Olson, [date]
    Communication skills in helping relationships : a framework for facilitating personal growth / Vonda Olson Long.
        p.   cm.
    Includes bibliographical references and index.
    ISBN 0-534-33869-0
    1. Counselor and client.   2. Interpersonal communication.
3. Counseling.   I. Title.
BF637.C6L66   1996
158' .3—dc20   95-8350
                                                  CIP

Dedicated to the travelers on the journey of growth, and to those with the desire and intention of helping fellow sojourners along the way—with a special tribute to those who already recognize that they can only help others travel where they themselves have gone.

A special dedication to my parents, Harold and Edna Olson, and my fellow sojourners Jack O'Neil, Wayne Maes, and Michael Wile, who have been unparalleled role models, teachers, and mentors on my own journey; and to Jeffrey, Katie, and Crystal, who have given me inspiration.

Contents

3

Purpose and Goals of the Helping Relationship: Facilitation of Growth toward Psychological Health   27

7

Helper Personal Growth: The Counselor as Client   87

P A R T

III

A Stage Structure for the Helping Process: Pre-Stage Attitudes of Rights, Respect, and Responsibility   95

8

Underlying Beliefs and Personal Growth: Rights, Self-Respect, and Self-Responsibility   97

9

Attitudinal Goals in Facilitating Personal Growth: Rights, Respect, and Responsibility   109

10

Rights versus Control   117

11

Respect versus Judgmentalness   125

12

Appropriate Responsibility
versus Rescuing and Blaming   133

15

Mechanisms of Communication: Verbal and Nonverbal Behavior, Statements, and Questions   167

16

A Five-Step Communication Model   177

P A R T

V

A Stage Structure for the Helping Process: Stages and Skills   191

17

Stage 1: The Presenting Problem   197

18

Stage 2: Underlying Issues   211

19

Stage 3: Direction and Change   233

Epilogue: The Framework in Review 245

References 249

Index 255

Preface

P eople are typically motivated to communicate by one of three
common reasons. The most common is communication to have
one's own perspective understood, by other people and by oneself.
The second is communication to try to understand *other* people's perspectives. Finally, the least common reason is communication to help
other people better understand their *own* perspectives. Although the last
is the type of communication least often used, it is probably the most
effective and productive, particularly in facilitating personal growth.

PURPOSE AND OBJECTIVES

The purpose of this book is to provide a framework for the helping process, which includes the identification of goals, a communication
model, and communication skills, for people who seek to help others
better understand themselves and their own perspectives. It is for
people whose goal in communicating is, at least in part, to help facilitate the self-understanding and personal growth of the person with
whom they are communicating. When the goal is to facilitate the other
person's self-understanding and personal growth, the interaction can
be considered a helping relationship, and the communication process
can be considered a helping process.

For the reader, objectives include these:

GENERAL OBJECTIVES
1. To be able to understand and apply an operational stage and skill-
   based framework for the helping relationship
2. To self-assess and develop skills relative to both self-development
   (facilitation of personal growth) and skill-development (facilitation of another's growth)

SPECIFIC OBJECTIVES
1. To be able to clarify your individual purpose and goals in communicating
2. To be able to identify a philosophy of human growth as applied to both yourself and others
3. To self-assess and develop communication skills effective for helping relationships
4. To self-assess and develop facilitative skills (effective use of communication skills)
5. To be able to apply an operational stage framework for growth facilitation in helping relationships

THE FRAMEWORK

This book presents a framework for the helping process. It includes both a system of outcome and process goals—what you want to accomplish—and a stage structure for the helping process—how you can accomplish it.

The system of outcome and process goals presents goals not only for the person with whom you are communicating (hereafter referred to as the client) but also for you (the helper or counselor) and for the helping relationship. These goals are presented as both outcome and process goals. An outcome goal is what you hope ultimately to accomplish. A process goal is the activity involved in accomplishing it. In this system, the client goal is psychological health, as evidenced by the development of a positive self-concept (outcome goal) through personal growth (process goal). The helper goal is effective helping (outcome goal) through growth facilitation (process goal). The helping relationship goal is the development of a catalytic (growth-promoting) relationship (outcome goal) through a growth-facilitating environment (process goal).

The three conscious dimensions of the human experience—affective experience (feeling), behavior (acting), and cognition (thinking)—are used to identify outcome goals for (1) client growth, (2) helper skills, and (3) the helping relationship. Consequently, the system of outcome and process goals presents goals for affective experience, behavior, and cognition for client, helper, and the helping relationship (see Preface Table).

The stage structure for the helping process is a guide for facilitating these goals. It includes a stage hierarchy, with pre-stage attitudes and three sequential developmental stages. The stages address the personal growth of the client; the helping skills of the helper, including a five-step communication model and corresponding helper skills; and the growth environment of the helping relationship (see Preface Table).

The foundation of this framework is three underlying beliefs and attitudes that are referred to as the three R's: (1) *rights*—individuals have the *right* to be and develop their unique selves, including feelings, behavior, and thoughts (short of harm to others); (2) *respect*—individuals deserve *respect* for their unique capabilities and differences; (3) *responsibility*—individuals have the right to, and corresponding responsibility for, their own actions, decisions, and choices.

PREREQUISITES

If one is to facilitate client growth and self-understanding effectively and intentionally, the following are prerequisites:

1. Clarification of purpose and goals
2. Philosophy of growth
   a. Self-understanding
   b. Other-understanding (empathy)
3. Structure for facilitation of growth
   a. Communication skills
   b. Facilitation skills (effective use of communication skills)

Key to effective helping is an organized, operational framework based on identified goals and incorporating a clear philosophy of growth as well as communication and facilitative skills. This book presents an example of such a framework. It emphasizes that self-understanding and the ability to apply one's philosophy to the facilitation of one's own growth is paramount to facilitating the growth of others.

ASSUMPTIONS, INTENTIONS, AND LIMITATIONS

The intention of the book is to provide one possible model of an operational, structural framework for the helping process. It does not purport to be the only possible framework that can be used in the helping relationship.

Eclecticism

The framework was developed for use with transitional, developmental, personal growth issues in a nonpathological population and is applicable to a wide range of helping relationships. It is presented as an eclectic structure. Rather than reflecting a particular theoretical orientation, it can accommodate a combination of theories. Pre-stage attitudes and Stage 1, for example, are particularly compatible with humanistic orientations; Stage 2 can embrace all counseling orientations, especially the cognitive approaches; Stage 3 is complementary to behavioral approaches.

Clarification of Goal Terminology

The goal of the helping relationship on which this framework is based is the facilitation of psychological health, a condition evidenced by and referred to with a wide range of terms: *self-fulfillment, self-actualization, self-realization, optimal development, full-functioning, mental health,* and a *positive self-concept.* For consistency, a *positive self-concept* is used in this book to represent the outcome goal of psychological health. Many other terms would be equally appropriate, and readers are invited to let *positive self-concept* represent whatever definition suits them best.

Diversity

The terms *psychological health* and *self-concept* carry varying connotations culturally. The reader is invited to consider them in the broadest possible sense. The terms are defined here as reflecting generally a *self-assessment of satisfaction with one's being and life.* Inherent in this definition is an implicit respect for individual values and diversity—whether measured by culture, ethnicity, race, language, family, gender-role, religion, ability/disability, socioeconomic status, education, sexual orientation, or any other means by which individuals define themselves.

The framework is a structure to organize the helping process. It does not presume to incorporate all necessary content for the development of an effective helper. There are entire books and courses on such major areas as counseling theory, multicultural perspectives, gender-role issues, ethics, and practicum issues. This book makes no attempt to address the content adequately from these major components of the helping relationship. Instead, such content, gained from other readings and coursework, may and should be infused into the framework.

Ethical Guidelines

Ethical standards, as outlined by the American Counseling Association (1988), the American Psychological Association (1981), and the National Board for Certified Counselors (1989), should be maintained at all times when using this framework in a professional helping relationship.

STRUCTURE

The book is divided into five parts. Part I addresses the purpose and goals of the helping relationship. It includes a discussion of outcome and process goals, identifying the goal of the helping relationship as client growth (process goal) toward psychological health, as evidenced by the development of a positive self-concept (outcome goal). A system of outcome and process goals for the client's personal growth, helper skills, and a growth environment for the helping relationship is presented in Part II. The stage structure for the helping process is presented in Parts III through V. Part III addresses the pre-stage attitudes of rights, respect, and responsibility (the 3 R's). Part IV addresses communication components and the five-step communication model. Part V presents the three-stage hierarchy and helper skills.

SUPPLEMENTAL WORKBOOK

A workbook to accompany the text, entitled *Facilitating Personal Growth in Self and Others,* is also available. It includes two parallel sections: Self-Development and Skill Development. The self-development section focuses on a sample philosophy and framework of personal growth to facilitate the development of psychological health through a positive self-concept. Since helper self-understanding is paramount to effective helping, the goal of this section is to facilitate helper self-understanding and personal growth. Workbook exercises in the self-development section, however, may also be used with clients. The skill development section focuses on helping skills and provides exercises for the development of effective attitudes, communication, and facilitation skills for promoting the personal growth of others.

ACKNOWLEDGMENTS

This book emerged over several years from my developing and teaching a course in communication skills for counselors-in-training at the University of New Mexico. I would like to acknowledge, therefore, the role my students played in providing incentive as well as direct and indirect feedback during the development of this framework. I am also indebted to my clients, who were the recipients of its principles. My special thanks to the manuscript reviewers, who offered valuable suggestions: Ron Bingham, Brigham Young University; Larry Burlew, Barry University; Diane Coursol, Mankato State University; Kenneth Davis, Villanova University; Murray Finley, Rhode Island College; Lena Hall, Nova Southeastern University; Anthony LoGiudice, Frostburg State University; Greta Slaton, Northeastern State University; Sue Stephenson, University of Wisconsin–Stout; Dennis Warner, Washington State University; Brian Wlazelek, Kutztown University of Pennsylvania; and John A. Worsley, Community College of Rhode Island.

I would like to thank the professional staff at Brooks/Cole for the wonderful collaborative adventure in the development and production of this book. I would like to especially thank Rusty Johnson for his belief and confidence in this project; his encouragement prompted its undertaking. Claire Verduin and Lisa Gebo were most helpful in their persevering support and assistance with its completion, and thanks to Marjorie Sanders for her skillful overseeing of its editing and final production.

Appreciation goes specifically to Julie Gellert-Ross, Brenda McGee, Chelly Weiss, Lynn McDonald, and Mosafar for their research and feedback assistance, and to Trish Stevens, Sheri Lesansee, and Keli Alark for their contribution to typing and table development.

Special appreciation goes to Mark Taylor for his wonderfully creative original cartoons and art work; and finally, a special thanks goes to Gayle Griffith for her limitless dedication and unfailing support in the preparation of this manuscript.

— Vonda Olson Long

PREFACE TABLE Framework for the Helping Relationship

| | | *Client Growth Goals* | *Helper Skills* | *Helping Relationship Goals* |
		A Positive Self-Concept	*Effective Helper*	*Catalytic Relationship*
A System of Outcome Process Goals	Outcome Goals			
	Affective Experience	Self-Acceptance Self-Esteem Self-Actualization	Role Model Catalyst Facilitator	Trust Understanding Change
	Behavioral	Congruence Competence Control	Genuineness Positive Regard Focusing: Empathy and Expression	Rapport Processing Directionality
	Cognitive	Rights Respect Responsibility (3 R's)	Rights Respect Responsibility	Rights Respect Responsibility
	Pre-Stage Attitudes	*Growth* Rights, Respect, Responsibility	*Growth Facilitation* Rights, Respect, Responsibility	*Growth Environment* Rights, Respect, Responsibility
A Stage Structure for the Helping Process	*Model Step*		*Skills*	
	Stage 1 Presenting Problem 1. Listen	Telling the Story Initial Awareness • Thoughts • Feelings • Behavior Initial Problem Identification	Attending	Rapport and Trust 3 R's Stage 1 Client Process Stage 1 Helper Process Skills
	2. Center		Genuineness Positive Regard Boundary Distinction	
	3. Empathize		Explicit Empathy Concreteness	
	Stage 2 Underlying Issues 4. Focus	Problem Differentiation • External Disappointment • Internal Issues Examining the Issues New Perspectives and Insight	Empathic: Implicit Empathy Confrontation Expressive: Self-Disclosure Immediacy	Processing and Understanding 3 R's Stage 1 and 2 Client Process Stage 1 and 2 Helper Process Skills
	Stage 3 Direction and Change 5. Directional Support	Direction Implementation Change	Directionality Implementational Support	Directionality and Change 3 R's Stage 1, 2, and 3 Client Process Stage 1, 2, and 3 Helper Process Skills

If you can't identify your purpose, and don't know what you're trying to accomplish, how will you know whether, and when, you accomplish it? Effective communication skills are built on a knowledge of what you're trying to do, why you're trying to do it, and how you plan to accomplish it.

"Cheshire cat," she began, rather timidly, "would you tell me, please, which way I ought to go from here?" "That depends a good deal on where you want to go to," said the cat. "I don't much care where . . ." said Alice. "Then it doesn't much matter which way you go," said the cat.

— LEWIS CARROLL, *ALICE IN WONDERLAND*

Purpose and Goals of
the Helping Relationship

A s a counseling supervisor, I've observed many counseling student interns meeting with their first clients for their first sessions. One such session was especially memorable. The student intern began the session with "Hi, how are you today?" "Oh, okay," the client responded. "So, how can I help you?" the intern continued. The client responded by explaining that her father had recently died, and that with the money he had left in trust for her, she felt she could finally get out of an emotionally abusive relationship and go back to school. She didn't know, however, what she wanted to choose as a major, nor what courses to take the first semester. To this, the student intern stated that it wasn't necessary to decide on a major until one's third year, and the client really didn't have much choice of courses the first semester anyway, because of the general requirements. He added that if she just registered for any of the general requirements, she'd be okay. He then asked if she had any other questions. When she said that she "guessed not," he was going to end the session after what had been 15 minutes. He explained later in supervision that it "didn't seem like she had any more questions." When asked what he viewed as the purpose of counseling, he said, "To answer her questions." The intern was clear about his goal, but that might not have been sufficient. We must be sure that our purposes are both well defined and *appropriate* to the situation.

THE IMPORTANCE OF GOALS

In order to determine whether you've accomplished a goal, you must first know what the goal is. When you register for driver's education the goal is clear: to learn to drive. Determining whether the goal has been accomplished is fairly straightforward: Can you now drive, as determined by passing a driving test? In the helping relationship, where the activity is an interactive process, it's even more important to know what you are trying to accomplish. If you don't know what you are trying to accomplish, how will you know when, and whether, you accomplish it?

OUTCOME AND PROCESS GOALS

Goals are aims or objectives. They are attainable, reachable, achievable. The achievement of a goal is recognizable. *Outcome goals* are destinations, ends, final results. Mastering a certain piano piece, driving to a

specific destination, talking to communicate a set of instructions, planning and giving a party, reaching maturity—all are examples of outcome goals. *Process goals* are aims or objectives focused on movement, motion, and activity. Process goals are focused on the process. Playing the piano for the enjoyment of playing, talking, enjoying a party, growing—these are examples of process goals. The focus is on the process or activity itself rather than the outcome. Going for a drive for the pleasure of driving rather than for reaching a specific destination is an example of a process goal. Process goals can be for the sake of the activity itself (e.g., running to enjoy running) or in conjunction with an outcome goal (e.g., running to get in shape).

PURPOSE AND GOALS

Goals, whether process or outcome, are attainable, accomplishable. Process goals equate to the journey. Outcome goals equate to the destination. *Purpose* is your reason for going, like setting a course in order to arrive at a destination. In other words, the destination is the outcome goal. The traveling itself, like traveling west, is the process. The purpose is why you are going.

The outcome goal, then, is *where* you're going, the process goal is the *going*, and the purpose is *why* you're going. Part I of this book addresses purpose and goals as they apply to the helping relationship. In the framework for the helping process presented here, the *outcome goal* is identified as improved psychological health, as evidenced by a positive self-concept; the *process goal* as growth; and the *purpose* of helping as the facilitation of the process and outcome goals—or the facilitation of growth (process goal) toward psychological health and a positive self-concept (outcome goal).

1

Counseling and Helping Relationships

The important thing is this:
To be ready at any moment
to sacrifice what you are
for what you could become.

◆— CHARLES DUBOIS

W hen I was 10, I went to summer camp. I stayed in a cabin with three other girls just a short way from a cabin with a sign that read "camp counselor." Curious, I stopped. I found a 15-year-old girl sitting in the counselor's chair. Having been to camp three previous summers, she seemed very mature, knowledgeable about camp policy and activities, and generally wise. This camp counselor is a good example of the wide range of activities the term *counseling* is used to describe.

Counseling is a type of helping relationship. A helping relationship is one in which the identified helper has the express and intentional purpose of assisting the person to be helped with his or her emotional, mental, physical, and/or spiritual well-being. *Professional* helping relationships are those that are skill-based and for which the helpers receive payment for their services. Whereas numerous professions have the potential for incorporating a helping relationship, such as teaching, social work, nursing, the ministry, therapeutic recreation, student advisement, and administrative roles, counseling is probably the most commonly identified specific type of helping relationship. In this book *counseling* is always used to refer to the specific professional counseling process, and *helping* is used in reference to the more general helping relationship, which includes the counseling relationship. This usage reflects the distinction that although not all helping relationships are professional counseling relationships, all effective counseling relationships are helping relationships. Basic principles of counseling, however, can be applied to all helping relationships.

COUNSELING: A SPECIFIC TYPE OF HELPING RELATIONSHIP

The range of definitions for *counseling* is wide. Some are quite general:

> Counseling is a term that may be used to describe a variety of activities. (Patterson & Eisenberg, 1983, p. 22)

> Counseling is an activity, an active process. (Egan, 1990, p. 7)

Others are more specific:

> This intense emotional experience involves learning new behaviors, participating in risk-taking activities, gaining understanding of one's own behavior and others' reactions to one's behavior and sharing of personal feelings through self-disclosure. (Sanders, Jones, & Sanders, 1987, p. 249)

> Counseling helps clients edit and enlarge their cognitive "maps" of the world through interpretation, information giving, modeling, and selective reinforcement. (Brammer, Shostrom, & Abrego, 1989, p. 85)

Some point out what counseling is *not:*

> Counseling is not a random encounter, as social conversation, an inquiry into pathology, an exercise in "labeling," or a veiled interrogation in which the client must establish his or her "suitability" for service. (Wells, 1982, p. 83)

Most attempt to define both the process and purpose of counseling as well as goals and outcomes:

> Counseling is a relationship, and a process, and is designed to help people make choices and solve problems. (George & Cristiani, 1990, p. 3)

> Counseling has been described as a process in which a trained helper deliberately intervenes in the life of a client to assist that person in resolving concerns so as to live more effectively. (Humes, 1987, p. 49)

For our purposes, counseling is defined by four major criteria: (1) process, (2) goals, (3) purpose, and (4) parameters.

&

Counseling is (1) an intentional, systematic and replicable, skill-based, interactive process with (2) the goal of client growth toward improved psychological health, (3) the purpose of facilitating that goal, and (4) specific focal, ethical, and logistical parameters.

&

Counseling Is a Process

"Process" is "a series of actions or operations conducing to an end" (*Webster's New Collegiate Dictionary,* 1981). Counseling is an intentional, systematic and replicable, skill-based interactive process.

Intentional

Although we learn many of life's lessons quite accidentally, counseling is intentional, or purposeful. It does not happen by chance, luck, or accident. We plan for the process to happen and actively facilitate it.

Systematic and Replicable

The counseling process is systematic. It occurs in a fairly predictable sequence: It has a beginning, middle, and end (Patterson & Eisenberg,

1983). It is characterized by anticipated movement toward identifiable goals. There is a recognizable structure or framework for the process. To have an intentional, planned, and actively facilitated process, we need a systematic structure from which to operate.

If a process is systematic, with a recognizable structure and framework, that process will also be replicable. The framework and structure of the counseling process can be and is repeated with different clients. This does *not* mean a replication of content, nor necessarily approach and techniques. It means that the *framework* or system from which you operate, including basic tenets and principles used to make decisions regarding approach and techniques, is replicable. If you have trouble replicating a structured approach, you probably need to reassess and clarify your framework.

Skill Based

Counseling is based on theoretical principles of growth and development. It applies helping skills. These are responses chosen with the intent of facilitating the growth of the client.

Interactive

The counseling process is interactive. The counselor and client interact or participate in verbal exchange with one another for a mutually agreed-on collaborative purpose. For most, this might seem obvious. Many clients, however, seek counseling expecting to explain their symptoms or problems and get a prescription—either medical or non-medical—for a cure.

Counseling is an interactive, collaborative process characterized by a unique relationship between counselor and client, leading to change in the client. Some have suggested that counseling occurs *only* when it is interactive—when both the client and counselor have worked during the session (Patterson & Eisenberg, 1983).

Counseling Has Identifiable Goals

Regardless of counseling orientation, general outcome goals for the counseling process can be identified as (1) improved self-concept or self-realization, (2) self-actualization, optimal development, full or effective functioning, and (3) psychological, emotional, or mental health.

General process goals can be identified as movement or progress *toward* these outcome goals.

Although clients will have different specific goals for the counseling process, such as preserving a marriage, losing weight, or gaining confidence, these will all fall under the umbrella of the general outcome and process goals. This is because the general goals have a direct relationship with common basic human needs (see Chapter 2).

Counseling Has a Distinct Purpose

The purpose of counseling is to facilitate the attainment of the counseling goals—that is, the client's growth process (process goal) toward improved self-concept, self-actualization, and psychological health (outcome goals).

Counseling Has Specific Parameters

There are many types of communication interactions: discussions, debates, arguments, conversations. Many include some of the same criteria that have been identified for counseling. Counseling, however, has three specific parameters that distinguish it from the others: focal, ethical, and logistical parameters.

1. *Focal Parameters.* The focus in counseling is on the need of the client. Unlike many other types of interactions, where there is a mutuality of sharing, counseling is focused solely on the growth and goals of the client.

   > [Counseling is] a situation where two people interact and try to come to an understanding of one another, with the specific goal of accomplishing something beneficial to the complaining person. (Bruch, 1981, p. 86)

   > [The counseling relationship is unique in that] it is the only human relationship in which two people make genuine two-way contact at a deep level of knowing and caring, and yet have as their shared aim the personal growth and enrichment of only one of the partners. (Moursund, 1990, p. 13)

Counseling and conversation are not the same. While conversation may employ similar communication and even specific helping skills (see Part V), counseling differs in that it focuses solely and singly on the client. Counseling uses helping skills for the specific purpose of facilitating the growth of the client. The following excerpts delineate the difference:

> [Counseling skills are similar to conversational skills, but differ] in that they focus on only one-half of the conversation dyad. Moreover, these foundation skills are goal-directed; they are not used to pass time pleasantly but to accomplish a specific purpose. They are used for the benefit of, and in the service of, the client. (Moursund, 1990, p. 13)

> Counseling is not the same thing as conversation. In conversation two or more people exchange information and ideas. The experience is usually casual and relaxed. People leave a conversation and move easily to other things. Counseling, on the other hand, is characterized by a much higher level of intensity. Ideas are developed more slowly, encountered at a deeper personal level, and considered more carefully. People leave a counseling experience mentally and emotionally depleted, yet still thinking about what was discussed. (Patterson & Eisenberg, 1983, p. 9)

> A conversationalist can monopolize the conversation with anecdotes about herself or can give up on an uninteresting partner and move on to someone else. The therapist, in contrast, must focus on the client who happens to be in her office. . . . While the client, of course, shares the responsibility for what happens, it is the therapist's job to ensure that the exchange is as therapeutic as the client will allow it to be. (Moursund, 1990, p. 3)

2. *Ethical Parameters.* Professional counseling and helping relationships require ethical parameters. "Offering to provide profes-

sional people-helping service obligates the helper . . . to function in an ethical manner" (Patterson & Eisenberg, 1983, p. 9). Professional counselors need to honor the Professional Code of Ethical Standards set up by the American Counseling Association (ACA, 1988), or the American Psychological Association (APA, 1981), or the National Board for Certified Counselors (1989).

3. *Logistical Parameters.* Counseling has specific logistical parameters, such as a designated meeting place, specific time frame, fee structure, referral system, emergency system, expectations, and goals.

In a professional counseling relationship, the designated meeting place should allow for privacy and confidentiality, including a reception area or waiting room and secured area for maintaining and storing client files. It should be quiet and comfortable, with a professional decor.

A specific time frame means a designated beginning and ending to the session at an agreed-on time during the week. Understandably, there will always be more to process than time allows during any given session; therefore, counselor and client must defer to the mutually accepted time frame.

Fee structures vary depending on the agency as well as the counselor's training, licensure, and market demand. Fee structures may be fixed or on a sliding scale, based on the client's ability to pay. Regardless of structure, fees should be clearly addressed and understood at the initial contact or inquiry.

A professional helper will face situations in which the client, after an initial contact or a lengthy helping relationship, needs professional help that the counselor is unable or unqualified to give. For this reason, the counselor needs to have a referral system network and be prepared to refer clients to other helping professionals when appropriate.

Professional counselors also need to have an emergency system in place so that their clients have access to help between sessions and during times when the counselors are out of town and unavailable. Emergency systems may incorporate treatment centers, professional colleagues, and/or 24-hour crisis hot lines.

Expectations and goals of the counseling sessions and helping relationship should be addressed and clarified on initial contact. If expectations and goals cannot be agreed on, professional helpers need to reconsider the extent to which they can be of help.

SUMMARY

A helping relationship is one in which the goal of the identified helper is to facilitate the growth of the person to be helped. Counseling is one example of a professional helping relationship. It is defined here as (1) an intentional, systematic and replicable, skill-based, interactive process with (2) the goal of client growth toward improved psychological health, (3) the purpose of facilitating that goal, and (4) specific focal, ethical, and logistical parameters.

2

Human Dimensions and Counseling Orientations

A ship is safe in harbor—
but that's not what ships are for.

— JOHN A. SHEDD

T he goal of the helping relationship is to facilitate psychological health and personal growth. If helpers are to achieve this goal, they must have an understanding of the basic human dimensions that make up the individual. These dimensions are addressed in the major counseling orientations; therefore, it is useful to examine such orientations for their potential contribution to the helping relationship. Knowledge of a number of counseling approaches is especially valuable with a framework for helping skills such as the one presented in this book, as the framework is designed to accommodate many different approaches from humanistic, cognitive, and behavioral orientations. It does not espouse a particular theory, but draws on fundamental principles primarily from these three orientations and can be used with a variety of theories.

Many different counseling theories have contributed to our understanding of the facilitation of psychological health: client-centered therapy, Gestalt, rational-emotive therapy, reality therapy, behavior modification, psychoanalysis. One way to organize the counseling theories used in this book is to divide them into four major orientations based on their primary intervention emphases, as shown here.

ORIENTATION:	EMPHASIS:
Humanistic	Affective experience (feelings, emotions, and existential "being")
Behavioral	Behavior (actions and "doing")
Cognitive	Cognition (thought processing, beliefs, and "choosing")
Psychodynamic	Unconscious (ego structure and personality restructuring)

The orientations are not limited to the identified emphasis but rather are used as the initial primary focus. No matter which emphasis you begin with, the others must ultimately come into play, since the human experience includes them all.

Dimensions of human experience include three basic conscious domains—affective experience, behavior, and cognition—that interact to create the basic human experience. Experience is affected by both the conscious and the unconscious.

Human experience is also affected by meta-level dimensions of metaphysical and spiritual domains. Although the helping skills presented in this book can be used with any major orientation, the focus is limited to the conscious domains of affective experience, behavior, and cognition (see Figure 2.1).

FIGURE 2.1   Basic human experience

Affective experience, in this framework, is considered to be inclusive of feelings, emotions, affect, and one's experience of one's essence or "being." Affective experience correlates most closely to the humanistic counseling orientation.

Behavior refers to actions and what one does, or "doing." Behavior correlates with the behavioral counseling orientation.

Cognition, in this framework, is inclusive of thoughts and beliefs, the thought process, or thinking, and the decision-making process, or "choosing." Cognition correlates with the cognitive counseling orientation.

Affective experience, behavior, cognition, and the unconscious dimensions, then, can be viewed as corresponding to these four major counseling orientations: humanistic, behavioral, cognitive, and psychodynamic (see Figure 2.2).

FIGURE 2.2   Human dimensions related to counseling orientations

HUMAN DIMENSIONS

The three domains of (1) affective experience, (2) behavior, and (3) cognition make up the basic conscious human experience. As the goal in helping is to affect the basic human experience, understanding some of the components and principles of these domains is important.

Affective Experience

According to *Webster's New Collegiate Dictionary* (1981), *affect* is "feeling; the conscious subjective aspect of emotion." *Feeling* is defined as the "expression of emotion; the character ascribed to something as a result of one's emotional state," and *emotion* is the "affective aspect of consciousness; state of feeling; the psychic and physical reaction subjectively experienced as strong feeling." Affect, then, refers to feelings and emotions that are separate from, yet integrally related to, human thought, behavior, and experience. Affect refers to feelings and emotions that accompany, underlie, or otherwise relate to clients' experiences and behaviors (Egan, 1990).

There are four primary emotions. Although some prefer to think in terms of love and fear as the two basic human emotions (Williamson, 1992), for our purposes, four emotions are most useful: mad, glad, sad, and scared. These serve as general common denominators, with the acknowledgment that a wide range of specific emotions fall under each one.

1. *Mad* refers to anger, covering a range from frustration to irritation to rage. It includes both guilt and resentment, and corresponding defensiveness and aggressiveness. Generally, anger results when an image or expectation that we have is not realized; we've made a judgment about it and assigned blame to ourselves or others. Guilt or defensiveness is anger directed at ourselves for what we did or did not do. The reward for guilt, defensiveness, resentment, and aggressiveness is the general feeling of being "right."

2. *Glad* includes love, joy, esteem, hope, pleasure, enjoyment, competence, and happiness. Temporary pleasure can be experienced from external rewards, such as receiving a present or award, but the deeper levels of gladness reflected in joy and happiness relate to one's experience of having lived up to one's *own* image or expectations of oneself.

3. *Sad* includes loneliness, hurt, grief, pain, disappointment, discouragement, and depression. Transitional depression is a low point, relative to other higher points. A mountain lake, for example, is high relative to sea level, but low relative to the surrounding moun-

tain peaks. Sadness is generally related to either external loss, such as the loss of a friend, or the internal loss of not having lived up to one's own image or expectation, such as not having accomplished enough, or not having done a good job.

4. *Scared* refers to fear and can range from doubt and anxiety to terror. Rational fear, such as fear of a coiled rattlesnake ready to strike, is fear of something with real, life-threatening potential. Such fear is a helpful signal telling you to pay attention, and it summons extra energy. Irrational fear, such as fear of rejection, is dread of something with an unrealistic or non–life-threatening consequence. Irrational fear can result in unnecessary avoidance, pain, and immobilization. It generally is based on an illusion—what we imagine might happen to us if we don't live up to our image or expectations.

Behavior

Behavior refers to "acting, functioning, or conducting oneself" in a particular way (*Webster's New Collegiate Dictionary,* 1981). Behavior is one's action or response to a stimulus and the environment. Behavior, in this framework, refers to actions and "doing."

Behavior is promoted by thinking and feeling. Behavior is the interaction, or interactive result of thinking and feeling (Hutchins & Cole, 1992). Events are interpreted through thinking: interpretations promote feelings; interpretations together with feelings promote behavior.

Behavior is both verbal and nonverbal. Behavior can include both action and nonaction, and verbal and nonverbal behavior. Verbal and nonverbal behavior are discussed in Part IV of this book.

Cognition

Cognition, according to *Webster's New Collegiate Dictionary* (1981), is the "knowing or process of knowing," including awareness, judgment, and beliefs, accessed through thoughts and reasoning power. Cognition includes both content (thoughts) and process (thinking). Content is the substance of thoughts—e.g., the belief that people are basically good—and process is the action of thinking—e.g., ruminating about one's anxiety over giving a talk. Cognition, used here, refers to thoughts, beliefs, thought process, decision making, and "choosing."

Nothing is good or bad, but thinking makes it so.

— WILLIAM SHAKESPEARE, *HAMLET*

Thoughts have positive power. A fundamental power of thought is that it is the first step in any accomplishment. Before you are able to do something, you must first hold the idea or thought in your mind. This principle is the basis for numerous psychological concepts and approaches, such as the self-fulfilling prophecy, positive thinking, and psychocybernetics. The idea is simply that if you *believe* something will happen, it more likely *will* happen. If you focus on tripping as you walk across the stage, you are more likely to trip. If you believe you're going to catch a fly baseball, you're more likely to catch it.

We'll see it when we believe it.

— WAYNE DYER

This principle was never more obvious to me than when I was learning to hang glide. My instructor stressed that, when landing, I *must* focus on the clearing where I want to land, *not* the barn or cactus I want to miss. Sure enough, I discovered that if I focused on the cactus, I'd land on it. You go to the place on which you are focused.

❧

We go where we focus.   ❧

The power of thoughts can also be problematic. Obviously, if our thoughts and beliefs influence what happens, then our thoughts, depending on what they are, can be a help or a hindrance. Positive thinking—"I *can* do it"—can be helpful, and negative thinking—"I *can't* do it"—can be a hindrance.

Irrational beliefs can be problematic. A basic premise of Albert Ellis's (1989) rational-emotive therapy is that when we develop "irrational beliefs," or beliefs that are illogical—e.g., I must have the approval of everybody in order to be an acceptable person—they interfere with our sense of well-being and happiness. The power of thoughts, when they are irrational, can be problematic. Faulty assumptions can be problematic. Lazarus (in Cormier & Cormier, 1991) describes three faulty assumptions that are problematic beliefs: (1) the tyranny of "shoulds"—what we assume we "must" do or be; (2) perfectionism—assuming we must be perfect; and (3) external attribution—assuming that sources outside ourselves have power over us.

Thoughts precede feelings and behavior. Individuals interpret and respond to the same event differently. Some students might view getting a "C" in a course as traumatic, some might view it as simply an accurate assessment of their work, and others might find it to be a symbolic liberation from having to maintain a 4.0 average. One's interpretation (thoughts and beliefs) of the event influences emotional responses (mad, glad, sad, scared) and behavioral action (studying harder, changing majors, celebrating) (see Figure 2.3).

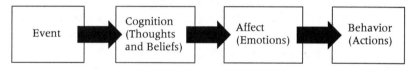

FIGURE 2.3   Behavior influenced by interpretation of event

Experience

The interaction of one's thoughts, feelings, and behaviors creates one's conscious human experience. Cognition creates the idea. Affect puts the idea into process. Behavior allows it to be accomplished. Cognition, affective experience, and behavior together make up our conscious experience, which is reflected in our being, doing, and choosing (see Figure 2.4).

External versus Internal Experience

External experience is what happens, and what happens to us. Internal experience is our perception and interpretation of these events.

FIGURE 2.4   Basic dimensions of conscious human experience

Unconscious and Meta-Level Experience

Human experience includes the impact of the unconscious and meta-level experience, such as spirituality and metaphysical dimensions. The focus and scope of this book, however, is limited to the conscious, basic human experience consisting of cognition, affective experience, and behavior (Figure 2.4).

COUNSELING ORIENTATIONS

The counseling process attempts to influence human experience. Counseling orientations, therefore, would logically correspond to the dimensions of human experience: affective experience, behavior, cognition, and the unconscious. The human being is the central core of the human experience; therefore, it should not matter which counseling orientation one uses as the orientation will ultimately impact all the dimensions and total experience of the human being.

Counseling Orientation by Human Dimension

HUMAN DIMENSION: Cognition
COUNSELING ORIENTATION: Cognitive
Primary Focus: Cognitive Human Experience—Thinking and thoughts, beliefs, decision making, and *choosing*.

Perspective Description: Cognitive therapy is based on a theory of personality maintaining that one's thoughts and thought processes determine one's feelings and behavior. Feelings are viewed as a direct result of thoughts rather than events. Therefore, if thinking changes, then changes will also take place in how one feels and behaves.

Counselor Role: To help the client investigate, reality test, and change problematic thoughts and thought processes.

Goal: Promotion of change in the client's thoughts and thought processes.

HUMAN DIMENSION: Affective Experience
COUNSELING ORIENTATION: Humanistic
Primary Focus: Affective Human Experience—Feelings and emotions, experience of essence, or *being.*

Perspective Description: The humanistic orientation includes approaches that focus on the affective human experience. Two founders of humanistic theory, Abraham Maslow and Carl Rogers, emphasized the dignity and worth of the individual, the person's ability to grow and change, and his or her freedom of choice and personal responsibility for behavior. Humanism stresses that given the satisfaction of basic physical, safety, love and acceptance, and esteem needs, humans possess an innate tendency and motivation to grow toward self-actualization and authentic existence.

Counselor Role: To help the client experience basic needs through role modeling an authentic relationship and providing a therapeutic environment.

Goal: Client self-understanding, self-realization, and self-actualization.

HUMAN DIMENSION: Behavior
COUNSELING ORIENTATION: Behavioral
Primary Focus: Behavioral Human Experience—Actions, behavior, and *doing.*

Perspective Description: Behavioral approaches are based on the belief that all behavior is learned, either by association, reinforcement, lack of reinforcement, or observation. If behavior is learned, it can be unlearned and relearned.

Counselor Role: To help the client identify problematic behavior, create conditions for learning and relearning, and develop strategies for implementing new behavior.

Goal: Behavior change toward more adaptive behavior, elimination of problematic behavior, acquisition of new behaviors and strengthening of desired behaviors.

HUMAN DIMENSION: Unconscious
COUNSELING ORIENTATION: Psychodynamic
Primary Focus: Unconscious Human Experience—Ego development and restructuring.
Perspective Description: Originated by Sigmund Freud, psychodynamic theory assumes that behavior is the result of unconscious motives. Anxiety arises from conflict among motives, and individuals attempt to control unacceptable motives by developing defense mechanisms, such as repression, denial, rationalization, intellectualization, and projection. Inability to control anxiety results in neurotic behavior (Carter, Van Ardel, & Robb, 1985).
Counselor Role: To help clients become more aware of the unconscious aspects of their personalities, work through unresolved developmental stages, and resolve unconscious conflicts through the use of free association, dream analysis, and analyses of transference and resistance.
Goal: Resolution of unconscious conflicts and integration of unconscious aspects of personality.

Orientations: Equal Effectiveness for Common Goal

Regardless of counseling orientation, the final goal of all counseling must be the same. All orientations must ultimately deal with the *whole* person—all human dimensions—in order to effect change in the individual's totality and ability to live more productively. Consequently, the dimension that is initially emphasized is not of major significance to counseling effectiveness. Indeed, research has indicated that counseling orientations are of fairly equal effectiveness (Osipow & Betz, 1991). The effectiveness of counseling appears to be more directly related to characteristics and skills of the counselor and to client readiness than to counseling orientation (Moursund, 1990). "It has often been asserted that *good* therapists, no matter how widely their theoretical frameworks may differ, seem to have about the same level of effectiveness" (Moursund, 1990, p. 12).

Orientation Commonality of Purpose and Process

Counseling orientations are all variations of a fundamental process with a common goal. They must, therefore, share basic commonalities in purpose, overall goals, core elements, and skills. Counseling is

> a generic process that includes essentially the same elements whether performed in a community counseling clinic, a rehabilitation center, a school, a hospital, or any other facility. While certain methods may result in more effective work with certain clients . . . the basic structure of the helping process is the same. (Patterson & Eisenberg, 1983, p. 2)

Good counselors, whatever their orientation, all

> share some important *skills*. They know how to listen, to attend to what a client is saying and to let the client know they are attending. They know how to help the client deal with problems, clarifying what the problem is, considering various solutions, making decisions. They know how to encourage the expression of feelings, and they have some guidelines about what to do when a strong feeling is expressed. (Moursund, 1990, p. 12)

Some have suggested that we cut across orientations to identify a set of "converging themes," or principles, approaches, and methods that constitute "the essence of helping" (Egan, 1990, p. 14).

Counselor-Client-Orientation Compatibility

Individual clients differ, and the human dimensions that may be most prominent in any given situation will vary. Consequently, therapies often appropriately involve more than one approach or orientation. Such a flexible approach allows counselors to adjust to their individual clients by using a variety of approaches that can address all human dimensions. "Because human behavior is complex, no one approach has the complete answer, nor is capable of helping all people in all situations" (Humes, 1987, p. 63).

It is also important for the counselor to fit well philosophically with the counseling orientation. The counselor must believe in the process in order for it to be effective.

ะ&

To try to engage in an ill-fitting counseling approach is like trying to run a marathon in shoes that don't fit.

ะ&

SUMMARY

Psychological health and growth, the identified general goals of the helping relationship, include at least the basic conscious human dimensions of affective experience, behavior, and cognition. The helping relationship, therefore, needs to acknowledge these basic human dimensions. They are reflected in the major counseling orientations. The humanistic orientation is identified with affective experience and being; the behavioral orientation with behavior and doing; the cognitive orientation with cognition and thinking; and the psychodynamic orientation with the unconscious. It is, therefore, useful to consider the major orientations and their potential contribution in the helping relationship.

3

Purpose and Goals of the Helping Relationship: Facilitation of Growth toward Psychological Health

If I am not for myself,
Who will be for me?
And if I am only for myself,
What am I?
And if not now, when?

— HILLEL,
SAYINGS OF THE FATHERS

G rowth and change in the client, leading to greater psychological health, are the desired end results of the counseling relationship; but we don't arrive there by accident. We must set out a defined procedure for achieving this end.

GOALS OF THE HELPING RELATIONSHIP: PSYCHOLOGICAL HEALTH AND GROWTH

You will not know whether you are being effective and successful in your actions unless, first, you are clear about what you are trying to

27

accomplish, and second, your behavior reflects your goals. Three typical problems can occur with the goals of the helping relationship:

1. *Undefined goals.* Undefined goals are vague or unclear. You will have difficulty recognizing success if you're not sure what you are trying to do.

2. *Inconsistent theory and practice.* The helpers state theoretical goals, such as client self-responsibility, that are inconsistent with their practice—for example, giving advice, wanting clients to do what the helpers tell them to do.

3. *Conflicted professional and personal goals.* Professionally, the helper understands that the client may have to experience discomfort in looking at an issue—perhaps remembering and revisiting a painful experience—but personally, he or she wants to manipulate the situation so the client will feel better—by avoiding or rescuing. The underlying and perhaps unconscious personal goal is really for the *helper* to feel better.

Goals need to be (1) clear, (2) consistent in theory and practice, and (3) professionally based. They must be well understood and grounded in professional judgment, followed by behavior and facilitative helping skills that are consistent with those goals.

General versus Specific Goals

From the helper's perspective, the general structure, framework, and goals of the helping process will be the same, regardless of the diversity of clients and client issues. The general framework and goals are based on dimensions of human experience and principles of human growth and development. General goals, such as client self-realization, self-actualization, psychological health, effective functioning, optimal development, or positive self-concept, cut across and provide a common umbrella of goals among clients. The helper needs to have clarity regarding general goals of the helping relationship before ever meeting with a client.

Specific goals are those the client, or client and helper, identify that are specific to the client, situation, and time. Specific goals, such as losing weight or becoming more assertive, self-accepting, confident, or expressive, fall under the umbrella of the general goals of the helping process. Specific goals cannot be identified prior to meeting with the individual client.

Outcome versus Process Goals

Outcome goals are end results. Process goals are a means, an action, or an activity for attaining the outcome goals. Whether a goal is a process or an outcome goal depends on whether the focus is on the process or the end result.

The helping process is a means to an outcome goal for the client. Statements like "We had a good session" must ultimately translate into movement toward an outcome goal for the client. If the process is effective, the client should be experiencing more of the identified outcome goal than before the helping process took place. "Virtually every significant theory of counseling states that creating some kind of change toward growth in the client is the ultimate intended outcome of the counseling experience" (Patterson & Eisenberg, 1983, p. 28).

It is important to distinguish between outcome and process goals. Outcome goals are intended results of the helping process. They can be described in terms of general outcomes—positive self-concept—and specific outcomes—assertive and confident behavior. Outcome goals will be reflected in actual change in the client that will be manifested outside the helper's office and after the helping process has ended.

Process goals are the means—the activities, actions, and behaviors—in which one engages en route to accomplishing an outcome goal. The general process goal for the client can be viewed as the developmental process, or growth. Specific process goals can include "plans for events that take place during the counseling sessions and in the [helper's] office" (Patterson & Eisenberg, 1983, p. 28).

OUTCOME GOAL: PSYCHOLOGICAL HEALTH AND A POSITIVE SELF-CONCEPT

The outcome goal is the end result, the destination, for which we're striving. There are many different counseling approaches, but the destination is still improved psychological health.

The destination of counseling is described with a wide range of outcome goals. These can include psychological health, self-fulfillment, a positive self-concept, effective functioning, optimal development, full functioning, self-realization, self-responsibility, self-actualization, and mental or emotional health.

Different orientations emphasize different human dimensions. Ultimately, however, all human dimensions are impacted because the common denominator is the human being and human experience. The outcome goals of counseling reflect common component themes that parallel the human dimensions. These themes include (1) one's view of oneself (positive self-concept, self-realization, self-image) or *being;* (2) how one lives one's life (effective functioning, full functioning) or *doing;* and (3) one's internal judgment (self-responsibility, self-actualization) or *choosing.*

Positive Self-Concept

Historically, many approaches have emphasized an improved view of oneself as the goal of counseling. An improved self-image, feeling better about oneself, increased self-acceptance, permission to realize one's authentic identity, congruence of one's ideal and real identities all reflect the range of focus related to self-concept.

> All successful therapies implicitly or explicitly change the [client's] view of himself from a person who is overwhelmed by his symptoms and problems to one who can master them. (Frank, 1971, p. 357)

> [The goal is for clients to approach] the realization that [they] no longer need to fear what experience may hold, but can welcome it freely as part of [their] . . . developing self. (Rogers, 1961, p. 185)

> A common goal of counseling is that the client will improve his or her self-concept and come to think of himself or herself as a more competent, lovable, or worthy person. (Patterson & Eisenberg, 1983, p. 23)

> The aim is individualization, or the realization of the self. (Schultz, 1990, p. 108)

> [To have the person become] more congruent, less defensive, more realistic and objective in his or her perceptions, more effective in problem solving, and more accepting of others—in short, the individual's psychological health and adjustment is closer to the optimum. (Patterson, 1986, pp. 410–411)

. . . seeing and caring about self in a new way; and achieving congruence between client perception and reality of experiences. The focus is on the emotional experience of the client. (Corey, 1991, p. 115)

Effective Functioning

Helping clients learn how to live their lives in a fuller and more effective manner is another historical emphasis of counseling outcomes. Effective functioning refers to the ability to make decisions and act, or function, in a way conducive to health, that is, effectively.

Effective individuals have a good sense of immediate, intermediate, and long-term goals, and they direct their efforts in a meaningful way toward appropriate goals. Less effective individuals are unable to develop plans, or they may be unable to implement the plans they have made. (Doyle, 1992, p. 32)

. . . to maintain and enhance the self to become a fully functioning person. (Schultz, 1990, p. 349)

The ideal outcome of [counseling] is a fully functioning person. (Rogers, 1961, p. 185)

Individuals who are functioning effectively usually (1) satisfy their needs in appropriate ways, (2) deal with the stresses of life and their emotional reactions by using effective coping processes, (3) learn tasks that are appropriate to their developmental stage, (4) have worthwhile social interactions and interpersonal relationships, and (5) demonstrate other positive attributes. Individuals who are functioning less than optimally often manifest problems in one or more of these areas. (Doyle, 1992, p. 17)

The [criterion] of therapeutic progress . . . is not a question of social acceptability . . . as viewed by . . . some authority . . . but the patient's own awareness of . . . more effective functioning. (Perls, 1969, p. 12)

Self-Actualization/Self-Responsibility

Actualizing one's individual potential and assuming responsibility for oneself—one's thoughts, feelings, behaviors, and resulting consequences—has also been a major theme of counseling goals.

> Counseling is a way of helping the individual help himself . . . to meet . . . life problems more adequately, more independently, more responsibly than before. (Rogers & Wallen, 1946, pp. 5–6)

> The focus is on the potential for self-actualization. The attainable goals include an understanding of free choice and consequence of actions. (Corey, 1991, p. 115)

> Healthy individuals exhibit the full range of human desires and emotions and have learned to respond to those feelings in a responsible manner. (Doyle, 1992, pp. 32–33)

> The ultimate goal is the actualization of self. (Schultz, 1990, p. 349)

> . . . to continue to develop one's potential, to grow and expand as a person. The need to actualize oneself and realize one's potentialities is central to . . . growth. (Ryff, 1989, p. 1071)

Affective experience, behavior, and cognition are all dimensions of the human experience and are integrally connected by the human being. In the same way, the outcome goals of positive self-concept, effective functioning, and self-actualization/self-responsibility are all integrally connected and are synonymous in that they all represent and refer to psychological health. In this framework, in order to establish and simplify terminology, the outcome goal of psychological health is hereafter referred to as a positive self-concept.

In the structure of outcome goals within this framework (see Preface Table), the general goal for the client is identified as a positive self-concept (outcome goal) attained through personal growth (process goal). Because a positive self-concept encompasses all three dimensions of the conscious human experience (affective experience, behavior, and cognition), affective experience goals, behavioral goals, and cognitive goals are identified (see Preface Table) and are addressed in Part II.

PROCESS GOAL: GROWTH

While the outcome goal—a positive self-concept—is the ideal for which we are striving, the process goal is the activity involved in the striving. The process is the activity, the movement, the change.

Counseling orientations may vary in their emphasis whether it be affective experience, behavior, cognition, or the unconscious; they also vary in approach, techniques, and specific outcome goals. The process goal, however, is always the same. Across orientations, the common process goal is client growth and change. "Virtually every significant theory of counseling states that creating some kind of change toward growth in the client is the intended [goal] of the counseling experience" (Patterson & Eisenberg, 1983, p. 5).

The Process Goal of All Counseling Orientations: Growth and Change

This commonality across counseling therapies is expressed in different ways. In humanistic orientations, the focus is on innate growth and change in awareness, affect, and insight. It is on changing one's perception of one's own being and on assuming responsibility for oneself. This theme has persisted over several decades.

[Growth is a] process of movement in a direction which the human organism selects when it is inwardly free to move in any direction. (Rogers, 1961, p. 187)

The goal of Gestalt phenomenological exploration is awareness or insight. (Corsini & Wedding, 1989, p. 323)

[In client-centered therapy, the fully functioning person is one who is fully feeling and fully experiencing—one who no longer fears experience, but learns to welcome it] as part of his changing and developing self. (Rogers, 1961, p. 185)

[Basic to the existential-humanistic philosophy is the recognition of the value and dignity of all human beings, and the assuming of self-responsibility] for choice and acting intentionally in the world. (Ivey, Ivey, & Simek-Morgan, 1993, p. 286)

Behavioral orientations focus on changing actions and behavior. This has historically included extinguishing behavior such as stopping smoking, relearning behavior such as responding to a group by speaking rather than withdrawing, and learning new behavior.

[The process goal of behavioral counseling is] to attempt to help the client change dysfunctional behavior to more functional patterns, such as overcoming shyness, reducing debilitating anxiety, controlling counterproductive anger, or reducing interpersonal conflicts. (Patterson & Eisenberg, 1983, p. 5)

[Joseph Wolpe's systematic desensitization technique, one of the most widely practiced examples of behavioral therapy, is based on a principle of reciprocal inhibition to change behavior] If a response antagonistic to anxiety can be made to occur in the presence of anxiety-evoking stimuli so that it is accompanied by a . . . suppression of the anxiety responses, the . . . anxiety responses will be weakened. (Wolpe, 1958, p. 71)

[The process goal in behavioral counseling is] a process of defining maladaptive or distressful behaviors and then establishing procedures to ameliorate them. (Cottone, 1992, p. 166)

In cognitive orientations, the process goal is to change the content and/or the way one thinks—to change one's thoughts, thinking, and decision making.

In cognitive-behavioral therapy the cognitive structures we use to organize experience are our personal schemas. We learned these constructions from past experiences, and changing ineffective schemas is an important part of therapy. (Meichenbaum, 1991, p. 7)

Cognitive therapy changes maladaptive behavior by restructuring cognition—that is, by altering the assumptions and premises that underlie the distortions. (Burke, 1989, p. 161)

The process goal of rational-emotive therapy is to change one's beliefs. (adapted from Ellis, 1989)

The basic goal of rational-emotive therapy is to change a self-defeating outlook on life and acquire a rational and tolerant thought process. (Corey, 1991, p. 208)

The immediate goal is to change the information-processing apparatus to a more "neutral" condition so that events will be evaluated in a more balanced way. (Corsini & Wedding, 1989, p. 286)

Successful Adlerian intervention changes behavior by focusing on the client's thinking; changes in behavior result from changes in the goals the client sets. (Burke, 1989, p. 226)

In reality therapy, the process goal is the assuming of responsibility, and control of one's life, and the facing of consequences for one's actions—accepting reality. (adapted from Glasser, 1965)

In psychodynamic orientations, the process goal is change resulting in making the unconscious conscious.

. . . to gradually uncover unconscious material. (Corey, 1991, p. 45)

. . . to bring repressed memories, fears, and thoughts back to the level of conscious awareness. (Schultz, 1990, p. 68)

The process of psycho-analysis . . . is a process . . . in which the [client] can explore [repressed] . . . memories surrounding the emergence of symptomatic behavior. (Cottone, 1992, p. 101)

Just as the word unconscious can be used to summarize Freud's theory, so can . . . free association be used to summarize psycho-dynamic [process] . . . encouraging the client . . . to say anything that comes to mind. (Ivey, Ivey, & Simek-Morgan, 1993, p. 162)

Growth and Change: The Need for Definition

As just demonstrated, growth and change are common process goals across counseling orientations. Growth always involves change— whether in affective experience, behavior, cognition, or consciousness. Change, however, does not always produce growth. To avoid ambiguity, in this framework, the process goal of counseling is considered to be growth.

"Growth" is "a stage in the process, action, or manner of growing; progressive development; growing or developing into maturity" (*Webster's New Collegiate Dictionary,* 1981). Growth, as it relates to the counseling process goal, is defined here as follows:

ò&

Growth is the process of action and change resulting in a pro-gressive movement toward improved psychological health as evidenced by the development of a positive self-concept.

ò&

PURPOSE OF HELPING RELATIONSHIPS: THE FACILITATION OF GROWTH

Purpose means "intention, having an aim or meaning" (*Webster's New Collegiate Dictionary,,* 1981). Purpose is *why* you're doing what you're doing. The purpose of a helping relationship is to *facilitate* the accom-plishment of the identified goal. In this framework, the purpose of the helping relationship, then, is the facilitation of growth toward the de-velopment of a positive self-concept.

ଽ

The outcome goal is improved psychological health as evidenced by a positive self-concept.

The process goal is growth.

The purpose is the facilitation of growth toward the outcome goal.

ଽ

There may be variations in the terminology for growth, but the purpose of the helping relationship across orientations is generally described as helping the client grow.

The purpose of [helping] is generally [to promote] the improvement or enhancement of the client or his/her existence. (Schmidt, 1980, p. 30)

The ultimate purpose of the [helping] experience is to help the client achieve some kind of change that he or she will regard as satisfying. (Patterson & Eisenberg, 1983, p. 5)

The ultimate aim of any [helping] process is to help clients move in a positive direction to find self-satisfaction through living a meaningful and authentic life. (Doyle, 1992, p. 35)

The purpose of [helping], broadly conceived, is to [help] empower the client to cope with life situations, engage in growth-producing activity, and make effective decisions. (Patterson & Eisenberg, 1983, p. 1)

When an individual, couple, group, or family is experiencing difficulty handling developmental issues, transitions, interpersonal interactions, or particularly challenging situations, an effective helper can supplement and assist in the process.

The purpose of [helping] is to supplement and assist . . . to step in to help when the natural growth process is not handling things well enough. (Moursund, 1990, p. 2)

People seek the services of professional helpers—counselors, social workers, psychologists, and psychiatrists—when their capacities for responding to the demands of life are strained, when desired growth seems unattainable, and when important decisions elude resolution. Sometimes the person in need of help is urged or required to seek counseling by a third party—spouse, employer, parent, or teacher—who believes the individual is failing to manage some important aspect of life effectively. (Patterson & Eisenberg, 1983, p. 1)

As the intention of helping is to facilitate the process goal of client growth, the accomplishment of that aim can be measured through the clients' achieving the goal. This does not mean clients need to have already arrived at an ideal outcome, but that you have effectively engaged them in the journey—the growth process. Because the goals reflect internal judgment by the client, assessment of their accomplishment must be determined by and large by the client.

[The accomplishment of purpose is determined by] achieving valued outcomes, as defined by the client. (Egan, 1990, p. 7)

. . . the client's statement that he or she is reasonably content with the outcome of treatment, either because the behavior complained about has changed or because his or her evaluation of the behavior has changed so that he or she no longer perceives it as a significant problem. (Fisch, Weakland, & Segal, 1985, pp. 122–123)

The purpose of helping is to assist the client in developing a positive self-concept. Accomplishing this requires having a clearly defined process for facilitating growth. To be an effective helper, you will need clarity in the following:

1. Purpose and Goals (Part I)
2. Philosophy and Framework for Growth (Part II) including
   • Self-Understanding
   • Other-Understanding
3. Stage Structure for the Helping Process: Facilitation of Growth (Parts III, IV, and V) including

- Communication Skills (Parts III & IV)
- Facilitation Skills (Part V)

SUMMARY

Clarification of the purpose and goals of the helping relationship is essential to effectiveness. General goals of the helping relationship are constant; specific goals change with the client, situation, and time. Outcome goals are end results; process goals are a means, action, or activity for reaching the outcome goals.

The process goal in all counseling orientations has been historically identified as growth and change. General outcome goals have generally focused on psychological, emotional, and mental health, as evidenced by such terms as a *positive self-concept, effective functioning,* and *self-actualization.* In this framework, the process goal is growth, and the general outcome goal is psychological health, evidenced by a positive self-concept.

The client goal is personal growth toward psychological health and a positive self-concept. The goal for the helper is effective communication skills to facilitate client growth, using the helping relationship as a growth environment.

THE SEASONS OF FREEDOM

I was born to the playground
in summer.
To soft sky, warm sun
and good earth.

I was free to move
with the breezes.
Free of fear, clothing
and girth.

I was still on the playground
in autumn.
Through for fear I'd be harmed
I was clothed.

I wandered restricted
and wondered
at the need for constriction
I loathed.

I stood on the playground
in winter.
Deep under layers of snowsuit
and wools.

My soul cried out
in anger
As I peered out at the
playground through holes.

The playground in Spring
appeared dampened
When they came
To remove my coat.

They said I should play
but I couldn't
I felt stiff and withered
and cold.

Now the playground again
is in summer
I look, lightly clad,
and grown,

And remember the movement
of breezes
and begin to recall
my own.

— VONDA LONG

A System of Outcome and Process Goals for Client, Helper, and Helping Relationship

I n Part I, the purpose and goals of the helping relationship were addressed, with the process goal identified as client growth, the outcome goal as client psychological health, evidenced by the development of a positive self-concept, and the purpose as facilitating the achievement of those goals, or the facilitation of growth. Three dimensions were also identified as making up the basic human experience. These included affective experience (feeling, being), behavior (behaving, doing), and cognition (thinking, choosing).

In Part II, a system of outcome goals is presented for the client, helper, and helping relationship. It incorporates both general outcome goals and goals from each human dimension: affective experience, behavior, and cognition (see Part II Table). The importance of developing a clear system of outcome goals prior to considering the helping process itself cannot be overemphasized. The purpose and goals of the helping relationship have simply been identified as facilitating growth toward a positive self-concept. Until we know what we *mean* by these goals, however, how can we facilitate their accomplishment?

In the system presented in this section, the general outcome goal for the *client* is identified as a positive self-concept sought through the process goal of growth. Also discussed are affective experience, behavioral, and cognitive goals that reflect growth toward the general goal. Self-acceptance, self-esteem, and self-actualization are identified as affective experience goals reflecting a positive self-concept. Although these three attributes are difficult to measure directly, they can be inferred from behavior. Congruence, competence, and internal control are identified as behavioral goals related to the affective experience of developing a positive self-concept. Behaviors are impacted by our cognition (thoughts and beliefs), and three basic underlying beliefs regarding rights, self-respect, and self-responsibility affect our ability to be congruent, competent, and internally controlled. These are identified as cognitive goals.

The general outcome goal for the *helper* is to be an effective helper through the process goal of growth facilitation. Affective experience, behavioral, and cognitive goals are identified for the helper, reflecting the development of effective helping skills. An effective helper is described as a role model, catalyst, and facilitator (affective experience goals). Behavioral goals for the effective helper are identified as genuineness, positive regard, and the ability to focus, through empathy and expression. Cognitive goals are identified as underlying attitudes about rights, respect, and appropriate assumption of responsibility.

PART II TABLE   A System of Outcome Goals

Outcome Goals	Client Growth Goals: A Positive Self-Concept	Helper Skill Goals: Effective Helper	Helping Relationship Goals: Catalytic Relationship
Affective Experience Goals	Self-Acceptance Self-Esteem Self-Actualization	Role Model Catalyst Facilitator	Trust Understanding Change
Behavioral Goals	Congruence Competence Internal Control	Genuineness Positive Regard Focusing: Empathy Expression	Rapport Processing Directionality
Cognitive Goals	Underlying Beliefs: Rights Respect Responsibility	Attitudinal Goals: Rights Respect Responsibility	Acknowledged Beliefs: Rights Respect Responsibility

Finally, the general outcome goal for the *helping relationship* is to be a catalytic relationship, pursued through the process goal of an effective growth environment. Again, affective experience, behavioral, and cognitive goals are addressed for the relationship, reflecting the development of an effective growth environment. A catalytic relationship is described in terms of trust, understanding, and change (affective experience goals). Behavioral goals are identified as rapport, processing, and directionality. Cognitive goals are identified as acknowledged beliefs in rights, respect, and appropriate responsibility-taking.

4

Client Outcome Goals:
A System for Personal Growth

*In answer to the question, "What do you want
to be?" one third grader wrote: "I would like
to be myself. I tried to be other things, but I
always failed."*

— ANONYMOUS

I f a goal is to have purpose, we need to know not only what the
goal *is* but also what it *means*. What is *growth*? What is *psychological health* and a *positive self-concept*? We need to understand what we
mean by these terms and the process of their development before we
can help facilitate their achievement.

There is obviously not a single system for personal growth, any
more than there is a single theory of human development. The system
presented here is *one* example of such a framework. The point stressed
is that helpers, to be consistently effective, need a conceptual system of
growth and the facilitation of growth (helping) from which to operate.
Readers are invited to consider this system in the development of their
own philosophy of growth to be used in conjunction with their *own*
philosophy of helping and growth facilitation.

Also, the preferred terminology for the outcome goal (psychological
health, positive self-concept) will vary with different cultural, gender-
role, ethnic, and spiritual perspectives. This is acknowledged. The
reader is invited to view this system's outcome goal of psychological

health as evidenced by a positive self-concept from the broadest possible perspective. The terms are defined generally as satisfaction with oneself (one's essence or being) and how one is living one's life (personal choices, actions, living). Inherent in this definition is an individual, internal assessment of satisfaction with one's life. Individual clients determine whether they are satisfied with themselves, whether their lives are fulfilling, and to what extent there is a "problem" or "issue." The definition also acknowledges the incorporation of cultural, ethnic, gender-role, individual, and spiritual perspectives by the individual client. Additionally, readers may wish to replace terminology used here with words that better reflect their individual cultural, ethnic, and spiritual perspectives.

GROWTH

Growth is a "process of change moving toward an outcome; a gradual development toward full formation or optimal maturity." Maturity is the state of "being fully grown or fully developed" (*Webster's New Collegiate Dictionary*, 1981). Growth, then, is movement toward maturity. Maturity, in the context of the helping relationship, is a positive self-concept.

ॐ

Personal growth is the process of developing a positive self-concept.

ॐ

A POSITIVE SELF-CONCEPT

"Our self-concept is the conception or picture we have of ourselves. It is the way we see ourselves. It is what we think ourselves to be" (Arkoff, 1988, p. 3). A person's self-concept is initially formed through contact with others and is continually operating in relationship with others. Consequently, cultural, ethnic, familial, gender-role, racial, spiritual, and individual values become an inherent part of self-concept. Regardless of the infusion of a wide variety of differing values, however, a positive self-concept in its broadest sense means that individuals feel satisfied with the essence of who they are, in the context of their own lives and relationships, *given and including* the infusion of their own values.

A *positive* self-concept, simply put, means the individual has a *positive* conceptualization or view of himself or herself. More specifically, it involves

1. The person's perspective: an evaluation based on his or her judgment
2. The person's conceptualization of self: the way that individual views his or her total being, including all the human dimensions of affective experience, behavior, and cognition, and within the context of relationship
3. The person's view of himself or herself as positive: a worthy, adequate, and acceptable being

These three criteria are important components of the definition of a positive self-concept.

ℰ

A positive self-concept is a conceptualization of oneself—including the human dimensions of affective experience (being), behavior (doing), and cognition (choosing)—which, in the individual's judgment, is worthy and acceptable.

ℰ

Self-Perspective

Self-concept arises explicitly from the perspective of the *self*. This does not imply isolation from others but rather internal judgment—judgment and approval that can come only from inside. "Self-concept is the person's picture of the self and the self-evaluation of this picture" (adapted from Brammer, 1988).

Self-perspective is inherently *internal*, as distinguished from other-perspective, which is *external*. Whereas the perspective of others has an influence on an individual's self-concept—particularly initially—it is not the same as, nor can it substitute for, that person's *self*-concept. A positive self-concept is based on one's own *internal* perspective and judgment, and will reflect individual and cultural values.

Conceptualization of Self

Self-concept means one's conceptualization of one's total identity.

> By the self I mean the concept of the individual as articulated by the indigenous psychology of a particular cultural group, the shared understandings within a culture of "what it is to be human." (Cushman, 1990, p. 599)

> . . . distinguishing what is directly and immediately a part of oneself from what is external to oneself. The self-concept is the person's picture or image of what he or she is, and might like to be. (Schultz, 1990, p. 355)

In wanting approval and acceptance, we focus externally on cues and responses from others, trying to adjust who we are to gain the greatest possible positive reinforcement. In attempting this adjustment to meet external expectations, we often develop two selves: an "external self," developed in response to our attempt to meet external expectations and gain external approval, and an "internal self," based on our true experience of ourselves. When the gap between what we believe and how we behave becomes large, this discrepancy can make it difficult for us to know who we really are. We become, in essence, discrepant selves.

Conceptualization of the self refers to how a person sees his or her true identity. When there is a discrepancy because of conflict between internal and external control, the disruption affects one's ability to conceptualize himself or herself as an integrated total person.

Positivity

A positive self-concept means experiencing oneself positively—as worthy, good, and acceptable. It means feeling good about who you are, your essence, your being. A positive self-concept is "to experience love in ourselves" (Williamson, 1992, p. xviii).

ॐ

A positive self-concept is feeling good about yourself and how you're living your life.

ॐ

Perhaps the most significant aspect of this definition is that it relies on the *individual's* judgment. It requires *internal* approval. It involves the *person* approving of himself or herself. *External* approval, while it may be influential, cannot substitute for *internal* approval. Relying on internal approval does not mean disregarding the external perspective, but it sets the individual as the ultimate judge of himself or herself.

A positive self-concept then, requires, first, the recognition and discovery of oneself, or self-identity, including the integration of discordant beliefs and behavior, or discrepant selves; and second, the internal approval of that self.

If we are to comprehend the process of developing a positive self-concept, we must understand its components. The basic components of a positive self-concept in this system are the affective experience of self-acceptance, self-esteem, and self-actualization, together with the underlying behavioral dimensions of congruence, competence, and internal control, and underlying cognitive beliefs concerning rights, self-respect, and self-responsibility.

COMPONENTS OF A POSITIVE SELF-CONCEPT: AFFECTIVE EXPERIENCE

A positive self-concept incorporates three components of affective experience: self-acceptance, self-esteem, and self-actualization. Just as affective experience, behavior, and cognition are human dimensions affecting our total human experience, self-acceptance, self-esteem, and self-actualization are dimensions of our affective experience of ourselves (see Figure 4.1).

To have a positive self-concept, a person must first *have* a self-concept, or self-identity. Second, a person must have internal approval

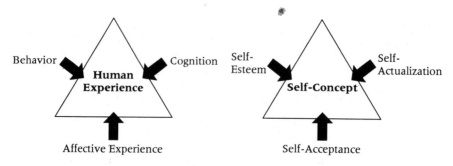

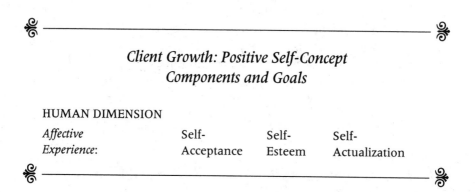

FIGURE 4.1   Components of human experience and self-concept

—a positive conceptualization—of himself or herself, including all the human dimensions: affective experience, behavior, and cognition. If the affective experience is positive, the individual will experience self-acceptance, self-esteem, and self-actualization. As components of a positive self-concept, these three attributes are outcome goals of the growth process. They constitute the *affective experience* to which the individual aspires. They are the results sought.

Client Growth: Positive Self-Concept Components and Goals

HUMAN DIMENSION			
Affective Experience:	Self-Acceptance	Self-Esteem	Self-Actualization

---

Self-Acceptance

Self-acceptance means that a person embraces who he or she is—in totality. Although a deceptively simple concept, it is the most challenging aspect of growth, and the most difficult self-perception to change (Hurley, 1989; Yahne & Long, 1988).

Definition

Acceptance means acknowledging and receiving something, just the way it is. Self-acceptance means acknowledging and receiving yourself, regardless of your perceived shortcomings and without critical judgment.

While self-acceptance is a part of a positive self-concept, it needs to be distinguished from it. A positive self-concept is feeling good about yourself. Self-acceptance is *valuing* yourself, despite shortcomings, which you may accept without valuing. Self-acceptance, then, *allows for* a positive self-concept:

> . . . individuals valuing and feeling good about themselves, in spite of their shortcomings. (Waite, Gansneder, & Rotella, 1990, p. 266)

> [. . . individuals realistically aware of their own strengths and weaknesses who feel their] unique worth. (Chebat & Picard, 1988, p. 355)

> . . . valuing oneself regardless of perceived shortcomings. (Shostrom, 1974, p. 18)

Two assumptions underlie self-acceptance: (1) you know yourself (self-identity)—something must first be identified before it can be accepted; and (2) you believe you have the *right* to be who you are in totality—feelings, thoughts, and behavior.

è❧

Self-acceptance is a person's unconditional acceptance of his or her total *being*, regardless of perceived shortcomings.

è❧

Self-acceptance is often confused with worthiness, judgment, agreement, and passivity. Self-acceptance is *not* the same as worthiness. Worthiness is a given: you have worth because you *are*. A human being has worth.

Acceptance is not judgment. Acceptance is unconditional and therefore requires suspended judgment. It means you accept reality for what it is.

Acceptance does not mean approval, consent, permission, agreement, or endorsement. You don't have to condone or even like what you accept; you simply acknowledge it. For example, I can *accept* the reality that I have straight hair, curly hair, diabetes, artistic talent, no musical talent, or a certain ancestry without necessarily *liking* it. I can *accept* someone's opinion without necessarily *agreeing* with it. Self-acceptance is also *not* the same as passivity or inaction. Just because I *accept* myself does not mean I am not actively working toward improving myself. For example, I might *accept* or acknowledge that I am a slow reader; at the same time, I may be taking a speed-reading course to improve my reading speed. Until we *accept* the reality of something, we can't move beyond it.

ॐ

Self-acceptance is *not* the same as worthiness, judgment, agreement, or passivity.

ॐ

Self-acceptance means you experience your self-identity as unconditionally acceptable. It means you are able to acknowledge the essence of your being. You are able to differentiate and separate your essence from your thoughts, feelings, and behaviors, and to experience that essence as acceptable. It means you may not like certain thoughts, feelings, or behaviors, but you can still experience yourself as worthy and acceptable.

Obstacles

External beliefs base our worthiness on external approval. When we are conditioned toward external approval and development of external beliefs, these can result in (1) lack of self-identity, (2) priority for doing rather than being, and (3) incongruence in true beliefs and behavior. All of these may create major obstacles to the development of self-acceptance. A lack of self-identity suggests you do not know who you are. It's hard to *accept* who you are before you *know* who you are. Prioritizing "doing" over "being" may result in *conditional* acceptance—

acceptance contingent on certain behaviors. Self-acceptance is *unconditional*. If you allow external approval to become too important, you develop a discrepancy between who you really are and who you appear to be. This can lead to the unhealthy belief that your internal real or true self is unacceptable.

When you believe that your innermost self is unacceptable, you may develop a mask in hopes of gaining external approval. If you use the mask as a defense to protect your real self from the pain of unacceptability, however, you will not only prevent the experience of acceptance for your true self, but almost surely guarantee a *lack* of self-acceptance.

&

The lack of self-identity, prioritizing of achievements, and development of an externalized self, or "mask," can create major obstacles to self-acceptance.

&

Self-Esteem

Self-esteem literally means "esteeming the self." It is often confused with self-worth. Self-worth is a given. You are worthy because you *are*. Self-worth (inherent) is not the same as self-esteem (the degree to which you value yourself).

Definition

Self-esteem is valuing and respecting yourself, based on your perceived strengths, attributes, worthwhile qualities, and actions. It is a positive regard for yourself based on your experience of yourself. Self-acceptance requires only that we accept ourselves regardless of short-comings; self-esteem requires perceived strengths that can be valued. Acceptance is given for simply *being;* esteem must be earned through *doing*.

> Self-esteem refers to feeling . . . positively about the ability of the self. (Løvlie, 1982, p. 13)

> Self-esteem is the ability to value one's self . . . a feeling represented through behavior. (Satir, 1988, p. 22)

> Self-esteem is valuing the self based upon perceived strengths. (Shostrom, 1974, p. 5)

> It is a personal judgment . . . regarding one's qualities as well as . . . behaviors. (Joubert, 1990, p. 1147)

Two assumptions underlie self-esteem: (1) Individuals know themselves (self-identity). "If I have not yet developed a self—I cannot esteem it! I cannot esteem what I do not have" (Løvlie, 1982, p. 13). (2) Individuals *respect* themselves and their actions.

❧

> Self-esteem is self-respect a person feels in recognition of his or her own perceived strengths, attributes, and actions.

❧

Self-esteem is based on actions. It comes from feeling good about how you're living your life. Actions are not the same as outcomes. Self-esteem is not based on wins or losses, but, to quote a cliché, on "how you play the game." There is no substitute for self-respect.

Obstacles

The development of external beliefs and the focus on external sources for approval may create major obstacles to the development of self-esteem. We learn to believe that external approval is more important than our own approval. Our worth becomes dependent on others' approval. We may learn to believe our esteem is relative to others' successes or failures and that external sources are responsible for our sense of value. As a result, we may develop obstacles, including (1) a false sense of self-esteem based on others' shortcomings rather than on our own strengths, (2) a readiness to blame others rather than develop our own competencies and abilities, and (3) a tendency to use achievements and accomplishments as a way to *prove* our worth and gain external approval.

Self-Actualization

Self-actualization, according to Abraham Maslow (1954), is an innate human need and motivation. It is the highest level of human psychological functioning and has been used synonymously with psychological health. It is a state of fulfilling one's potential. It is one of the most optimistic and life-affirming concepts in psychology (Daniels, 1988). Rogers (1961), in addressing this concept, defines it as "full functioning," and suggests it is "a process, not a state of being . . . a direction, not a destination" (p. 186). He sees

> the good life . . . [as] the process of movement in a direction which the human . . . selects when it is inwardly free to move in any direction, and the general qualities of this selected direction appear to have a certain universality. (p. 187)

Definition

Self-actualization means actualizing, or becoming and being, the person you truly are. It is a process of self-discovery, an ongoing unfolding of your true self. Finding this person means discovering your self-identity, realizing your innate potential, making congruent all the human dimensions—thoughts, feelings, and behaviors—and integrating the real versus the ideal aspects of your personality.

Self-actualization has been defined variously:

> . . . the full use and exploitation of [one's] talents, capacities, potentialities. (Maslow, cited in Mittleman, 1991, p. 115)

> The organism strives not only to maintain itself but also to enhance itself in the direction of wholeness, integration, completeness, and autonomy. (Brammer, 1988, p. 37)

> It is not betraying or suppressing one's own inner nature or potential. (adapted from Hall & Lindzey, 1985, p. 209)

Two assumptions underlie self-actualization: (1) The individual truly knows himself or herself (self-identity).

ॐ

> The individual alone can recognize the true actualization of his or her self.

ॐ

(2) The individual is *responsible* for his or her own actualization.

> In actualizing the self, you must accept responsibility for yourself, for your reality, and for making changes as necessary. (adapted from Craig, 1989)

> [Self-actualization is being] responsible for behaviors that affect the self. (Gladding, 1988, p. 157)

Self-actualization is the process of becoming oneself and actualizing one's potential.

Obstacles

Self-actualization has been described as the result of a natural tendency for humans to grow and develop to their fullest potential as their lower-level basic human needs are met (Maslow, 1954; Rogers, 1961). Self-actualization, as the highest level of human psychological functioning, can develop only after the more primary needs—(1) physiological, (2) safety, (3) love and belongingness, and (4) esteem—are met (Maslow, 1954). Given this prerequisite, one obvious obstacle to the development of self-actualization could be *a deficiency in satisfying lower-level needs.* "Self-actualization is reached when all [lower-level] needs are fulfilled" (Heylighen, 1992, p. 41).

As only the individual can recognize an accurate manifestation of his or her true identity, a second obstacle could be the focus on and *prioritization of external approval.* Only you can judge and give approval to yourself. Deference to external judgment for approval may be an obstacle to your highest self-actualization. "Self-actualization must be fundamentally independent of . . . socialization, or permission" (Daniels, 1988, p. 24).

Third, by believing others are responsible for our worth, happiness, and well-being, or correspondingly, to blame for our unhappiness, we may develop an *external locus of control.* We believe external sources have control over our lives. This belief could be a direct obstacle to self-actualization because to believe you have the ability to actualize yourself, you need to have control over yourself.

> Self-actualization is closely related to an internal locus of control.
> (adapted from Woody, Hansen, & Rossberg, 1989, p. 221)

Self-actualization means taking responsibility for achieving your full potential. It means investing in the process of becoming, of moving constantly to higher levels of growth and fulfillment. As Victor Frankl (1985) emphasized, the most powerful impact on our happiness is our own internal response and attitude toward what happens to us. "Individuals who are functioning effectively, accept responsibility for their actions" (Doyle, 1992, p. 33).

COMPONENTS OF A POSITIVE SELF-CONCEPT: BEHAVIORAL EXPERIENCE

Self-acceptance, self-esteem, and self-actualization are aspects of a positive self-concept that reflect outcome goals of a person's affective experience, or how that person feels about himself or herself. While affective experience is difficult to evaluate or affect directly, it can be inferred by behaviors associated with affective experience, which can be observed:

Congruence underlies self-acceptance. Self-acceptance is valuing oneself regardless of perceived shortcomings. Congruence—the consistency of one's thoughts, feelings, and behaviors, and integration of discrepant selves—reflects behaviorally such a valuing and acceptance of oneself. Congruence reflects a self-given permission of *being* oneself.

Competence underlies self-esteem. Self-esteem is valuing oneself based on perceived strengths. Competence reflects these strengths in observable and measurable behavior based on innate and developed abilities. It can be viewed as effective functioning. Competence is reflected in *doing*.

Internal control underlies self-actualization. Self-actualization is the process of self-development to the fullest. It requires internal control, the confidence of trusting your beliefs and judgment to guide your behavior, because only you can know what beliefs and actions are consistent with your innermost identity. Internal control shows that you are *choosing* who you will become and how you will live your life.

In this system, congruence, competence, and internal control are identified as underlying behaviors reflecting self-acceptance, self-esteem, and self-actualization.

Client Growth: Positive Self-Concept
Components and Goals

HUMAN
DIMENSION

Affective Experience:	Self-Acceptance	Self-Esteem	Self-Actualization
Behavior:	Congruence	Competence	Internal Control

Behavioral components can be observed, and therefore learned and developed. Developing congruence, competence, and internalized control is inherent in the growth process of developing a positive self-concept.

ॐ

Growth is developing the behavioral components of a positive self-concept: congruence, competence, and internal control.

ॐ

Congruence

Definition

Congruence means the "quality or state of agreeing, corresponding, or coinciding" (*Webster's New Collegiate Dictionary,* 1981). In the context of a positive self-concept, it means that your actions, speech, or other external behaviors reflect your internal thoughts and beliefs—not someone else's. Congruence is the integration of discrepant selves; it is

. . . consistency between ideal (perceived) and real (experienced) selves. (adapted from Woody, et al., 1989)

[When] words, actions, and feelings match—they are consistent. (Cormier & Cormier, 1991, p. 25)

You are *not* congruent when there is a discrepancy between what you believe and how you behave. Incongruence occurs when you behave so as to please someone else, even though doing so is contrary to what you believe and want to do.

ॐ

Congruence is the state of being true to yourself, reflecting agreement and consistency in your thoughts, feelings, and behaviors.

ॐ

Congruence emerges from the *belief* that you have the *right* to *be* who you are.

> We are most comfortable when we are being ourselves. We are upset when our thoughts and feelings and actions are not in harmony with the person we think we are . . . [or] when we are cast in situations or roles that demand behavior inconsistent with our concept of ourselves. (Arkoff, 1988, p. 1)

Congruence reflects the *affective experience of self-acceptance.*

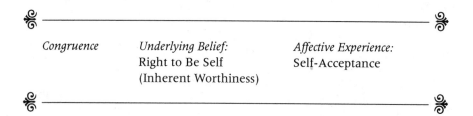

Congruence	*Underlying Belief:* Right to Be Self (Inherent Worthiness)	*Affective Experience:* Self-Acceptance

Obstacle: Fear of What Might Happen to Me/Fear of Rejection

A primary obstacle to congruence is the fear of what might happen to us if we let ourselves be who we are: the fear of rejection. Ironically, by not letting ourselves be who we are, *we* reject ourselves before others have a chance.

Competence

Definition

Competence means "having the qualities necessary for; sufficient for one's needs; sufficiency, capability, adequacy or ability" (*Webster's*

New Collegiate Dictionary, 1981). In the context of a positive self-concept, competence means that individuals develop and use their qualities and abilities to enhance, develop, and take care of themselves. It means using all their human dimensions in a way that reflects their adequacy and capability. It means living their lives in a way they can respect and feel good about. Competence involves the development and implementation of abilities and skills. It involves acting and doing. It is behaving capably, responsibly, and with integrity.

> According to Greenspan's model, personal competence includes physical competence, intellectual competence, and emotional competence. (McGrew & Bruininks, 1990, p. 54)

Competence means that individuals adequately use their capabilities to act with integrity in developing their potential and living their lives.

Competence is *not* depending on others to approve of and take care of you. There is no substitute for your ability to be responsible for yourself. Competence emerges from a *belief* in and *respect* for your own capability. Self-respect increases as we discover and develop our capabilities. A respect for our own capability frees us to take responsibility for ourselves. The affective experience of self-esteem reflects competence.

Self-esteem results from a self-respect for one's own competence—the ability to live one's life in a way one is proud of.

Competence	Underlying Beliefs: Respect for Self (Capability and Adequacy)	Affective Experience: Self-Esteem

Obstacle: Fear of What I Might Find Out about Myself/Fear of Failure

A primary obstacle to competence is the fear of what we might find out about ourselves if we try to do something we've never done before. That fear is the fear of failure. Ironically, by not trying, we not only get no practice building the skills we need, but we never get to experience the success necessary to reinforce a belief in our competence. We automatically fail.

Internal Control

Definition

Control means "to regulate, or to have power over" (*Webster's New Collegiate Dictionary*, 1981). Internal control, in the context of the self-concept, means an internal regulation of power over oneself, including all the dimensions. Internal control means we have a choice, we can make decisions, we can *think*.

乷

Power of choice carries with it both the freedom and the right to choose, and the burden of responsibility for our choices.   乷

Internal control is taking responsibility for your choices and taking charge of your life. Responsible action acknowledges both choice and accountability.

Efficiency is getting the job done right. Effectiveness is getting the right job done. (John-Roger & McWilliams, 1991, p. 281)

Internal control is cultivating individuality, accepting oneself, and defining one's own freedom and accepting responsibility for that freedom. (adapted from Hall & Lindzey, 1985)

Internal control focuses internally:

The only thing you can take charge of is the space within the skin of your own body. (John-Roger & McWilliams, 1991, p. 93)

Aside from a few biological necessities (eating, staying warm, sleep, basic needs), we don't *have* to do anything. You do what you do because of choice; to the extent you want to say you *have* to, you're focusing externally. (adapted from Conway, Vickers, & French, 1992)

You are *not* internalizing control when you are blaming, rescuing, or being irresponsible. You are *not* internalizing control when you hold external sources accountable for your life, happiness, thoughts, feelings, or actions.

> *If you don't like what you're doing, you can always pick up your needle and move to another groove.*
>
> — TIMOTHY LEARY

Internal control is acknowledging your right to choose and accepting responsibility for your choices.

Internal control emerges from the *belief* that you are *responsible* for yourself and how you live your life. The exercise of responsibility is the acknowledgment of choice and subsequent accountability. Internal control is reflected in the *affective experience* of *self-actualization*.

A self-actualizing individual must be exercising internal control.

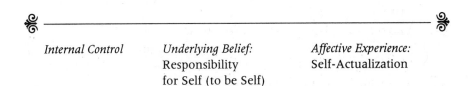

Internal Control	*Underlying Belief:* Responsibility for Self (to be Self)	*Affective Experience:* Self-Actualization

Obstacle: Fear of the Unknown/Fear of Abandonment

A primary obstacle to internalized control is the fear of the unknown. That fear is of loss and abandonment. We are conditioned to and familiar with an external locus of control: focusing externally, and depending on others and external sources to decide how we should be and how we should live our lives. To focus internally for such decisions is to pursue an unknown; it means leaving behind the familiar. In abandoning the familiar, we fear abandonment. Ironically, by not allowing ourselves to choose or be ourselves, we are abandoning ourselves.

> We stay in our comfort zone, the area of the familiar, because we think that being there will protect us from bad things, and doing something new takes us out of the comfort zone. But our safety is an illusion. Bad things happen regardless of where we are. (adapted from John-Roger & McWilliams, 1991, p. 259)

COMPONENTS OF A POSITIVE SELF-CONCEPT: COGNITIVE BELIEFS

Your beliefs about yourself have a powerful impact on your experience and behaviors. Underlying beliefs will affect your ability to accept, esteem, and actualize yourself; they will affect your ability to be congruent, competent, and internally controlled.

Specific beliefs affecting the behaviors of congruence, competence, and internal control, respectively, are identified as the *right* to be oneself, self-*respect* for one's capability, and self-*responsibility* for one's actions and choices.

In this system, these underlying beliefs regarding rights, respect, and responsibility (the three R's) lay the foundation, or pre-stage setting, for the helping process. The client's underlying beliefs regarding these three R's provide the cognitive goals for a positive self-concept, and helpers' underlying beliefs about themselves influence their *attitudes* toward their clients. Helper attitudes regarding the three R's, therefore, provide the foundation for helper skills and the helping relationship. These underlying beliefs and attitudes are addressed in Part III.

Outcome goals for client growth might be conceptualized as follows. The general outcome goal is a positive self-concept. The components or specific outcome goals are (1) the affective experience of self-acceptance, self-esteem, and self-actualization; (2) the underlying

Client Growth: Positive Self-Concept Components and Goals

HUMAN DIMENSION			
Affective Experience:	Self-Acceptance	Self-Esteem	Self-Actualization
Behaviors:	Congruence	Competence	Internal Control
Underlying Beliefs:	Rights	Respect	Responsibility

behaviors of congruence, competence, and internal control; and (3) the underlying cognitive beliefs of rights, self-respect, and self-responsibility (see Figure 4.2).

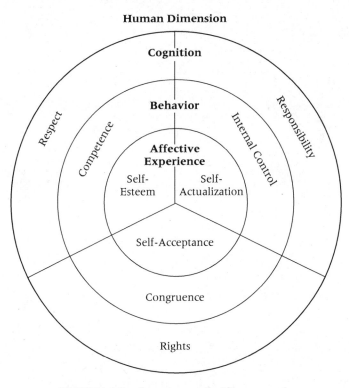

FIGURE 4.2   Outcome goals: Client growth

SUMMARY

In this system for personal growth, the client outcome goal is identified as improved psychological health, as evidenced by a positive self-concept. The process goal is growth toward that outcome. A positive self-concept is defined as a positive conceptualization of yourself, or feeling satisfied with yourself and how you're living your life.

The components of a positive self-concept can be viewed in terms of the human dimensions of affective experience, behavior, and cognition. Each dimension has three components, shown below.

Positive Self-Concept Components

Affective Experience:	Self-Acceptance	Self-Esteem	Self-Actualization
Behavior:	Congruence	Competence	Internal Control
Cognition (Underlying Beliefs):	Rights	Respect	Responsibility

---

5

Helper Outcome Goals: A System for Helping Skills

He not busy being born is busy dying.

— Bob Dylan

A system of goals for helping skills in facilitating client growth can be conceptualized using the same structure presented in Chapter 4. This is because the same human dimensions through which we can view human or client growth (affective experience, behavior, cognition) are present in helpers as they work with their clients. Thus we can conceptualize outcome goals for helper skills through (1) underlying cognitive beliefs, (2) behaviors, and (3) affective experience (see Figure 5.1).

UNDERLYING BELIEFS AND ATTITUDINAL GOALS

Your underlying beliefs about yourself get translated into your perspective on others. The underlying beliefs regarding (1) your right to be yourself, (2) self-respect for your capabilities, and (3) self-responsibility for life choices and happiness that you apply to yourself are translated into your *attitudes* toward others. If helpers believe they have the right to be themselves, have respect for their own capabilities, and take responsibility to make their own choices, these beliefs should translate into an *attitude* toward their clients. The attitude would likely include

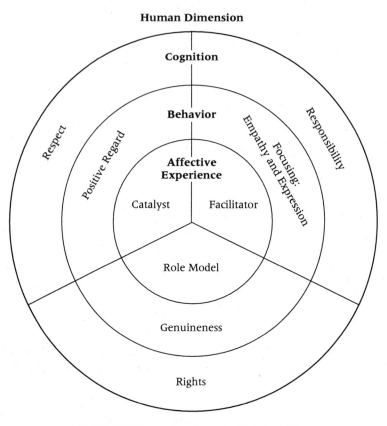

FIGURE 5.1   Outcome goals: Helper skills

(1) the belief that their clients have the right to be themselves, (2) a re-spect for their clients' capabilities, and (3) the belief that their clients are responsible for their own life choices and actions (see Chapter 9).

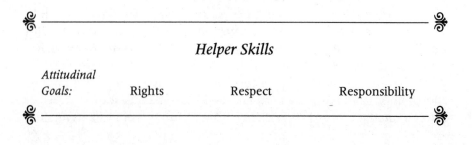

Helper Skills

Attitudinal Goals: Rights   Respect   Responsibility

BEHAVIORAL GOALS

The underlying beliefs regarding your own rights, self-respect, and self-responsibility, together with the resulting attitudinal beliefs regarding the rights, respect, and responsibility of your clients, provide the foundation for counselor behavioral goals of (1) genuineness, (2) positive regard, and (3) focusing through empathy and expression.

Helper Skills

Attitudinal Goals:	Rights	Respect	Responsibility
Behavioral Goals:	Genuineness	Positive Regard	Focusing: Empathy and Expression

---

Genuineness

The underlying belief and attitude that all individuals (including helper and client) have the right to be themselves encourages helpers to be congruent with themselves, reflecting an experience of self-acceptance and genuineness with their clients. Genuineness means being authentic, real, and sincere with another person.

While genuineness and congruence are related and sometimes used interchangeably to refer to an integration or concordance of internal awareness and external communication, they are differentiated here. Congruence refers to the internal state of consistency people experience when their beliefs, feelings, and behavior match.

[Congruence means] words, actions and feelings match—they are consistent. (Cormier & Cormier, 1991, p. 25)

Genuineness, in contrast, refers to the external manifestation of congruence. It is the individual being spontaneous, open, and natural—that

is, congruent—in the presence of another person. While one can be congruent by oneself, one needs another person present in order to be genuine.

> Counselors who are genuine do not hide behind a mask and play a role. (George & Cristiani, 1990, p. 129)

> Genuineness . . . is *acting without using a facade,* functioning openly without hiding behind the veneer of one's role. (Belkin, 1988, p. 71)

In the helping relationship, genuineness means being real, authentic, and congruent within the relationship with your client.

> Genuineness simply means that the counselor is being herself or himself in the relationship, avoiding presenting a facade or acting with contrivance because she or he is the therapist. [It] refers to communicating a realness in relationship. (Cormier & Cormier, 1991, p. 214)

Genuine people are at home with themselves and therefore can comfortably be themselves in their interactions. (Egan, 1990, p. 69)

It means avoiding the temptation to hide behind a mask of professionalism. (Corsini & Wedding, 1989, p. 172)

Positive Regard

The underlying belief and attitude that one is worthy of self-respect and the corresponding attitude of respect for others' capability to be self-responsible allows the helper to treat the client with positive regard. Positive regard means accepting other people's right to their own unique individualities and perspectives. It means respecting individuals' abilities and right to make individual choices regarding their own lives and happiness, short of harm to others. It means respecting others' perspectives without judging the individuals for holding these views.

Positive regard is a necessary antecedent to empathy—the understanding of another's experience or perspective—as we cannot begin to understand something we are not first willing to acknowledge and accept. Positive regard, in the context of the helping relationship, means actively communicating a judgment-free acceptance and respect for the clients' feelings, perspective, and experience.

This is not the same as agreeing with, condoning, or encouraging decisions or behaviors. The helper does not need to agree with clients' behaviors in order to communicate positive regard for the clients as individual persons with a right to their own perspectives and lives.

Focusing: Empathy and Expression

The underlying belief and attitude that one has the *right* to be oneself, together with the *respect* for one's capability to make choices necessary to be true to oneself, result in the recognition that individuals must be *responsible* for their own lives and choices (see Part III). This attitude regarding self-responsibility allows helpers to assume a role of facilitating *focus* without assuming responsibility *for* their clients. Just as focusing in photography involves getting a clear image, in the context of a helping relationship it means helping clients see clearly their issues, challenges, direction, and goals.

Focusing can incorporate approaches that are both empathic (communicating an understanding of the client's perspective) and expressive (communicating an understanding of the helper's perspective). The process of focusing and the behavioral goals of positive regard and genuineness are addressed in Part V.

AFFECTIVE EXPERIENCE AND OUTCOME GOALS

The attitudinal goals regarding rights, respect, and responsibility provide a foundation for the counselor's behavioral goals of genuineness, positive regard, and focusing. These attitudes and behaviors together promote outcome goals *for the helper.* Outcome goals for the helper, in the context of this framework, include being an effective role model, catalyst, and facilitator.

Helper Skills: An Effective Helper

Attitudinal Goals:	Rights	Respect	Responsibility
Behavioral Goals:	Genuineness	Positive Regard	Focusing
Affective Experience:	Role Model	Catalyst	Facilitator

Role Model

Observation and role modeling provide one of our most powerful and salient learning processes. A role model is one who serves as an example of a particular behavioral role for another to emulate. A positive role model is generally one who is unusually effective or inspiring in a particular role and so serves as an example for others. In role modeling, a person learns through observing and imitating the behavior of another person. Role modeling involves observation, imitation, and practice.

The helper is in a position to be a powerful role model. The clients' goal will be to find greater satisfaction in their own lives, to improve their functioning, to feel better about themselves, or generally to be happier with themselves and their lives. Since they perceive the counselor to have expertise in this area, they will quite naturally view him or her as a role model for the changes they are seeking.

> The client automatically models and patterns the attitudes, values, beliefs, and behaviors of [the counselor]. (Corey, 1991, p. 297)

> When therapists present themselves as real persons who make mistakes now and then, who are uninterested in prestige, who can laugh at themselves, who care for others, and who exhibit an optimistic courage, [clients] can begin to see that they, too can be like this. (Phares, 1992, p. 320)

> Clients tend to do what the helper does. (adapted from Brammer, 1988)

> As clients experience the counselor listening to them in an uncritical accepting way, they gradually learn how to listen acceptingly to themselves. (adapted from Corey, 1991)

Most counseling approaches recognize, if not actively seek to use, the impact of helper role modeling. In the context of this framework, helpers who communicate a belief and attitude that clients have the right to be themselves, and also demonstrate congruence and genuineness, serve as role models for these attitudes and behaviors.

> The counselor affects client change by modeling various healthy attributes such as congruence, empathy, genuineness, and self-respect. (adapted from Rogers, 1980)

> One of the counselor's main functions is to serve as a behavioral model for the client. This modeling may be explicit, as in role-playing activities, or implicit, as in direct and appropriate self-expression. (adapted from Mikulas, 1978)

By demonstrating genuine recognition and acceptance of his or her own failures and shortcomings within the therapeutic encounter, the counselor serves as a nonperfectionistic model for client self-acceptance. (May, 1989, p. 143)

Catalyst

In addition to being a role model, the helper serves as a catalyst for change in the client. A catalyst is something that precipitates a process of change without being changed itself (*Webster's New Collegiate Dictionary,* 1981). In the helping relationship, counselors promote change in the client without necessarily being changed themselves.

Bugental (1987), Rogers (1980), and Yalom (1980) all argue that a true therapeutic encounter is characterized by inner change in both the client and the therapist as a result of the relationship. They claim that when the primary therapeutic instrument is oneself, it seems unlikely that one could simply be a catalyst, emerging from the process no different from the way one was on entering it (Bugental, 1987). Concerning the catalytic role of the helper, however, change in the *helper* is not a prerequisite for change in the client. The helper may be impacted, deeply touched, or in a variety of ways affected by the helping relationship, but this is not *essential* to client change.

In the context of this framework, the helper's attitude of respect and communication of positive regard for the client can be viewed as catalytic for client growth. Respect and positive regard are consistently described as powerful catalysts for growth.

The counselor's communication of total acceptance and regard of clients as worthwhile persons frees them to accept and explore themselves. (adapted from Rogers, 1958)

Facilitator

As role models, helpers, through being themselves, are demonstrating attitudes and behaviors. As catalysts, counselors show attitudes of respect and positive regard that allow clients the freedom to grow. In addition to acting as role models and catalysts, counselors also serve as *facilitators*. A facilitator is one who makes a process easier (*Webster's New Collegiate Dictionary,* 1981). In the helping relationship, facilitation of growth means the active, intentional use of interventions designed to promote client growth.

While helpers as catalysts promote change simply through their presence, their *being,* and their attitude, counselors as facilitators are actively and intentionally *doing,* with the purpose of promoting change. Facilitators are consciously, thoughtfully, and deliberately using skills to promote growth.

In the context of this system, the behavior of focusing, which is promoting the development of self-responsibility, reflects the counselor as facilitator. Focusing, and specific facilitative skills associated with it, are discussed in Parts IV and V.

The facilitation of client growth toward the overall outcome goal of a positive self-concept can be approached by addressing and facilitating the development of (1) underlying beliefs or attitudinal goals (rights, self-respect, and self-responsibility), (2) behavioral goals (congruence, competence, and internal control), or (3) affective experience (self-acceptance, self-esteem, and self-actualization).

These approaches generally correspond to the major counseling orientations discussed in Part I. Addressing the underlying cognitive beliefs reflects a cognitive orientation, focusing on behavioral goals corresponds to a behavioral approach, and addressing the affective experience reflects a humanistic orientation.

An alternative conceptualization of the facilitation of client growth could be viewed as addressing the underlying belief, behavior, and affective experience of a number of human dimensions: being/feeling, choosing/thinking, or doing/behaving. The dimension of being/feeling, which would correspond to a humanistic orientation, would focus on the underlying belief of one's right to be oneself, the resulting behavior of congruence, and the reflective affective experience of self-acceptance. The dimension of doing/behaving (behavioral orientation) would focus on one's underlying belief regarding self-respect and capability, the related behavior of competence, and affective experience of self-esteem. The choosing/thinking dimension (cognitive orientation) would address one's underlying belief regarding self-responsibility, resulting behavior of internal control, and affective experience of self-actualization.

SUMMARY

In this conceptual framework for helping skills, the helper outcome goal is to be an effective helper, using effective helping skills. The process goal is effective growth facilitation.

The components of effective helping can be viewed in terms of the human dimensions of affective experience, behavior, and cognition; they are summarized below.

Effective Helper Components

Affective Experience:	Role Model	Catalyst	Facilitator
Behavior:	Genuineness	Positive Regard	Focusing
Cognition (Attitudes):	Rights	Respect	Responsibility

6

Helping Relationship
Outcome Goals:
A System for a
Growth Environment

God comforts the disturbed
and disturbs
the comfortable.

◆— UNKNOWN

I n Chapters 4 and 5, a system was presented for looking at outcome goals of client growth and helping skills. It was based on (1)cognitive beliefs and attitudinal goals, (2) behavioral goals, and (3) affective experience and outcome goals. In this chapter the same system is used to examine outcome goals for the helping relationship (see Figure 6.1).

COGNITIVE BELIEFS

One's underlying beliefs translate into attitudes toward others. In the helping relationship, these beliefs and attitudes affect client-helper interactions. The relationship needs to provide an effective contextual environment for the facilitation of client growth. In this system, an important foundation for an effective environment is acknowledgment of

77

the cognitive beliefs regarding rights, respect, and responsibility—particularly the belief that one has the right to be oneself with one's own thoughts, feelings, and actions.

Helping Relationship Goal: Effective Environment

Acknowledged Beliefs:	Rights	Respect	Responsibility

BEHAVIORAL GOALS

The acknowledged cognitive beliefs of rights, respect, and responsibility, as they apply to both client and counselor (see Part III), provide a foundation for the behavioral goals of (1) rapport, (2) processing, and (3) directionality in the relationship.

Helping Relationship Goals: Effective Environment

Acknowledged Beliefs:	Rights	Respect	Responsibility
Behavioral Goals:	Rapport	Processing	Directionality

Rapport

A helping relationship needs some level of rapport to be effective. A basic tenet of this system is that effective communication of the cognitive beliefs (underlying belief and corresponding attitude toward others) regarding *rights, respect,* and *responsibility* will promote rapport between helper and client.

Rapport develops through an interactional process in which the following are present:

1. A perceived warmth and caring

> Good . . . therapists work to build rapport, lessen interpersonal anxiety in the relationship, increase trust, and build an interpersonal climate in which clients can openly discuss and work on their problems. Clients . . . need to feel cared for, attended to, understood, and genuinely worked with if successful therapy is to continue. (Deffenbacher, 1985, p. 262)

2. A perceived safety and acceptance

> Rapport is a tangible sense of contact and trust between the helper and the one being helped; it develops out of acceptance. (adapted from Moursund, 1990)

> In the presence of effective helpers, clients quickly sense that it is all right to risk sharing their concerns and feelings openly. . . . Nothing "bad" will happen as a consequence of sharing and there is a very real chance that something gainful will come of it. (Patterson & Eisenberg, 1983, p. 13)

3. A perceived trustworthiness and credibility

> From the initial contact, then, counselors must be perceived as trustworthy. (George & Cristiani, 1990, p. 124)

> Effective helpers inspire feelings of trust, credibility, and confidence from people they help. (Patterson & Eisenberg, 1983), p. 13)

Processing

The helping relationship is built on the behavior of processing. Processing means talking about the issues and looking at different perspectives for the purpose of gaining greater understanding and clarity. One "processes" in order to express feelings, gain insight, and determine future behaviors. Processing is done in the hope that one will learn, grow, and feel more complete, content, accepting, and at peace with whatever is being processed. In many respects, helping *is* processing.

In this system processing occurs within the context of the underlying belief and attitude of respect. Respect reflects the helper's belief that the individual client is capable, able to recognize personal values, and able to make decisions consistent with them.

Directionality

The acknowledged cognitive belief of self-responsibility must necessarily affect directionality within the counselor-client relationship. If both helper and client believe and acknowledge that the individual not only has the right but the responsibility for making choices and decisions regarding his or her own life, then within the relationship, decisions that are appropriate to the client *must* be distinguished from those appropriate to the counselor, thus setting the direction within the relationship. If you truly believe that the only person who can recognize whether he or she is happy is that person, then you must believe that he or she is the only one who can take responsibility for movement toward that recognition.

If individuals take responsibility for themselves and are in the process of identifying values and making decisions regarding their lives, a sense of direction and progress will emerge. This sense of direction, progress, and movement can be referred to as directionality (Egan, 1990) and is addressed in Part V.

AFFECTIVE EXPERIENCE AND OUTCOME GOAL

The outcome goal for the helping relationship is to provide an effective environment for the promotion of client growth. The behavioral goals of rapport, processing, and directionality provide a foundation for the outcome goals of trust, understanding, and change.

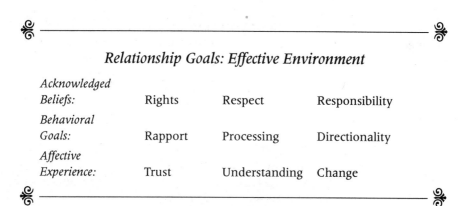

Relationship Goals: Effective Environment

Acknowledged Beliefs:	Rights	Respect	Responsibility
Behavioral Goals:	Rapport	Processing	Directionality
Affective Experience:	Trust	Understanding	Change

Trust

Trust is an affective experience and a desired outcome within the helping relationship. It grows from rapport and is built on acknowledged cognitive beliefs, particularly the belief that individuals have the right to be themselves and to have their own feelings, thoughts, and actions.

Trust is a prerequisite for a therapeutic climate.

> From the initial contact the helper must be perceived as trustworthy. (George & Cristiani, 1990, p. 124)

> Trust . . . refers to the ability to risk yourself, to put yourself in the hands of another. . . . [It] involves prediction—a prediction that if you put yourself in the hands of another, the result will be to your advantage. (Verderber & Verderber, 1989, p. 130)

> Trust is essential in order for the client to be open and revealing of very personal problems and concerns. (Cormier & Cormier, 1991, p. 53)

Trust is interactive. It involves another person. It also involves three elements: (1) *risk,* which results in loss or gain; (2) an outcome that is *dependent,* at least in part, on the other person; and (3) a sense of *relative confidence* that the other person will respond in a way that will result in gain rather than loss.

Trust cannot be mandated or forced. It can only be invited and developed. To encourage the emergence of trust, helpers can (1) become trustworthy, (2) develop self-trust, and (3) demonstrate appropriate trust.

Become Trustworthy

Trust is developed through trustworthiness. Trust involves a willingness to risk (i.e., being open and self-disclosing); *trustworthiness* involves a willingness to respond in a way that results in gain for the person trusting (i.e., response of acceptance, and support).

Clients trust helpers when the clients tell them personal information. The risk involves potential gain (being able to talk with someone; experiencing support, growth, and insight) and potential loss (loss of confidentiality if the helper gives the information to others). The gain or loss is in part dependent on the helper.

Helpers demonstrate trustworthiness when they honor confidentiality, are clear with their clients about the limits of confidentiality, and are supportive, accepting, and nonjudgmental. Helpers demonstrate trustworthiness by communicating an attitude of rights—that clients have a right to be who they are; by demonstrating respect—that the client's uniqueness, differences, and capabilities are worthy of positive regard; and by acknowledging appropriate responsibility-taking—that clients have both the right to and responsibility for their own decisions and choices.

Develop Self-Trust

If helpers don't trust themselves, how can they expect anyone else to trust them? If they want others to trust them, they need to be trustworthy to themselves. Self-trust also refers to confidence in their ability to deal with a situation in which their trust in someone else is disconfirmed. Competent judgments regarding *when* to trust is also part of self-trust. Generally, the greater the risk and potential loss (negative consequences), the greater the assessed trustworthiness needs to be.

Demonstrate Appropriate Trust

Appropriate trust is trust *based on* an assessment of trustworthiness. The risk needs to be balanced with the potential consequences and assessed trustworthiness. The greater the risk, the greater is the trustwor-

THE CLIENT MUST BE ABLE TO TRUST THAT HELPERS CAN HANDLE THEIR OWN EMOTIONS.

thiness. The lower the risk, the less need there is for trustworthiness. For example, giving someone a quarter and trusting her to bring you a cup of hot chocolate from a vending machine is a low risk because the potential negative consequence is simply the loss of a quarter. Assessment of trustworthiness becomes less important in this case. Lending a new car to a friend, sharing personal information about something that has caused you shame, or entering a joint business venture may be examples of higher risks. The potential negative consequences are greater; therefore, the assessed trustworthiness needs to be greater.

Understanding

Understanding means "the power to make experience intelligible" (*Webster's New Collegiate Dictionary,* 1981). The ability to find clarity and understanding of one's experience is the affective outcome goal of the helping relationship. The reason the helper and client engage in processing is to arrive at an understanding of the client's experience.

Both helper and client hope that understanding will result from the behavior of processing and believe that the client is capable and able to understand. In other words, respect for the client's ability to understand provides the reason for processing.

Change

Change—be it change of feelings, thoughts, or behavior—is the ultimate goal of the counselor-client relationship. The belief of self-responsibility lays the foundation for the behavioral goal of directionality, which, in turn, provides the groundwork for the promotion of change (see Figure 6.1).

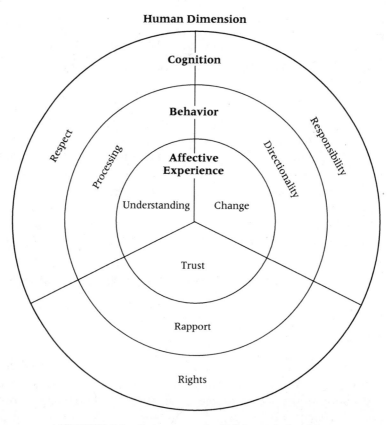

FIGURE 6.1   Outcome goals: Helping relationship

SUMMARY

In this system, the helping relationship outcome goal is an effective growth environment. The components of an effective relationship can be viewed in terms of the human dimensions of affective experience, behavior, and cognition, and are summarized below.

Effective Helping Relationship

Affective Experience:	Trust	Understanding	Change
Behavior:	Rapport	Processing	Directionality
Cognition (Acknowledged Beliefs):	Rights	Respect	Responsibility

---

7

Helper Personal Growth:
The Counselor as Client

*You cannot transcend
what you do not know.
To go beyond yourself,
you must know yourself.*

— SRI NISARGADATTA MAHARAJ

To suggest that you are knowledgeable and skilled at facilitating the growth of a fellow human being should mean that you are growing yourself, because you, too, are a human being. When I first moved to Albuquerque, I decided I wanted to learn to hang glide. After doing a little research, I discovered I had a choice of two instructors: one who was currently breaking records in distance and flying time competition off 11,000-foot Sandia Peak, and one who was in the hospital following a flying accident. It took only a moment to decide who I wanted as a flying instructor.

Generally, we look for guidance and instruction to those who are already demonstrating an ability to perform the task we want to learn. We believe that before people can help someone else do something, they need to know how to do it themselves.

In a helping relationship, if you are to help facilitate growth toward a positive self-concept for a client, you must be able to facilitate your own growth and be on the journey toward your own positive self-concept. It does not mean you must already have all your own personal issues completely resolved. Indeed, if that were the case, there wouldn't be anyone available to help anyone else. It does mean, however, that you have a knowledge and understanding of human growth and an ability to use that knowledge for facilitating human growth toward psychological health and a positive self-concept.

The importance of helpers' personal growth to their effectiveness cannot be overemphasized. Three areas specific to this importance are (1) the parallel process of growth facilitation, (2) the importance of boundary distinction, and (3) the helper as role model.

THE PARALLEL PROCESS: COUNSELOR GROWTH AS A PREREQUISITE TO HELPING OTHERS GROW

The knowledge, understanding, and philosophical approach for facilitating human growth toward a positive self-concept and psychological health is the same for the counselor as for the client. The approach is to facilitate the growth of human beings. Counselor and client are both human beings. The conceptual knowledge base that is applied to the client must, therefore, be the same for the counselor.

The important difference is that counselors are not distracted or focused on their own growth during the helping interaction, but in their own lives they have been and continue to apply the same principles of growth and learning to themselves that they are applying to their clients. Just as in the medical profession one learns to give injections by practicing on oneself, counselors have first applied the principles of growth to their own growth process and are well on their way to their own positive self-concept.

> Counselors must have positive self-concepts and feel secure about themselves [in order to help others effectively]. (Doyle, 1992, p. 6)

> Counseling is only as effective as the therapist is living effectively. (Belkin, 1988, p. 64)

> Counselors who have good feelings of their own self-worth, adequacy, and self-discipline transcend their own limitations and are free to give the necessary attention to their clients and to focus on ways to assist them. (Doyle, 1992, p. 6)

Counselors must accept themselves before they can accept clients sufficiently well to help them. (Brammer, 1988, p. 88)

Practice what you preach.

How can you possibly help someone else do what you are not able to do yourself? In counseling, where the purpose is to facilitate client growth toward a positive self-concept, you must be able to do for yourself—facilitate your own growth as reflected in your own positive self-concept—what you propose to help your client do.

BOUNDARY DISTINCTION

Boundary distinction refers to the ability to differentiate clearly between self and others, distinguishing personal issues from others' issues. You must be engaged in your own growth process in order to be prepared to help facilitate the growth of others; however, since the helping interaction is a time to focus solely on the needs and goals of the client, you must also be able to distinguish between personal issues and others' issues to be an effective helper.

Boundary distinction includes (1) self-awareness, (2) appropriate responsibility-taking, and (3) identification of potential counselor-client relationship issues.

Self-Awareness

The first step in effective boundary distinction as well as effective helping is counselor self-awareness. Without a knowledge of "self" it is impossible to distinguish "self" from "other."

Know thyself.

— (ANCIENT GREEK PROVERB)

Understanding clients is not enough. It is essential to understand your own assumptions, beliefs, values, standards, skills, strength,

weaknesses . . . and ways in which these permeate your interactions with your clients. (Egan, 1990, p. 25)

You should be aware of your own feelings, attitudes, values, and motivations for working with others. (Doyle, 1992, p. 6)

Self-awareness includes a recognition and identification of personal issues. Counselor personal issues, such as a need to be liked (including by clients), a belief that to be competent you must take responsibility for everything (including what might be appropriately your clients' responsibility), and difficulty dealing with anger (including your clients') can obviously interfere with your effectiveness in working with clients. When unrecognized personal issues result in such behaviors as agreeing with your clients (to gain their approval), telling them what to do (to "fix" the situation), or reassuring and agreeing to their demands (to appease their anger), the counselor's effectiveness is diminished. "Therapists must be able to look at their [clients] with objectivity and not become entangled in their personal dynamics" (Phares, 1992, p. 320).

Appropriate Responsibility-Taking

In addition to being able to distinguish between personal issues and client issues, effective helpers need to recognize what is their responsibility and what is their client's responsibility. This is a major issue of effective helping and is discussed in detail in Part III.

Identification of Potential Counselor-Client Relationship Issues

Several potential counselor-client relationship issues relate to boundary distinction and warrant specific recognition. Examples include dependency, projection, and performance anxiety.

Dependency issues refer to the counselor's using the counseling relationship to satisfy his or her personal needs at the expense of the client's needs. For example, when counselors recommend that the client stay in therapy longer than needed in order to satisfy the helpers' own needs (to be needed, to feel successful, etc.), there is a dependency issue. "Effective helpers like and respect themselves and do not use the people they are trying to help to satisfy their own needs" (Patterson & Eisenberg, 1983, p. 14).

Projection, in this context, refers to the counselor's making assumptions about the client, based not on an objective observation of the client's reality but rather on the *counselor's* perspective. For example, if you had angry feelings toward your ex-partner when *you* got a divorce, you *assume* your clients have angry feelings toward their ex-partners.

Projection can include feelings, needs, thoughts, opinions, beliefs, expectations, rules, and "shoulds." If I have certain feelings, opinions, rules, or beliefs, I can assume you do, too. If I look at my client and think "She shouldn't get angry," or "Why doesn't he just *do* it?" I am reflecting feelings and beliefs I hold about or for myself. "In order to avoid a dependent relationship with a helpee, it's important for you to be aware of your own needs, feelings, and problems . . . self-awareness enables you to communicate better . . . without . . . projecting your own feelings and needs" (Okun, 1992, p. 78).

Performance anxiety reflects a specific personal need of the counselor: the need to do well, be helpful, be competent, be liked. Counselors desire to be helpful, but if they are focusing more on their need to be helpful than on the clients' needs, paradoxically, they cannot *be* helpful. Clients, to be helped, need for you to focus on them and their needs, not on your desire to help them—which is your need.

Beginning counselors, in their desire to be helpful, often think too much about where the session is going, what to say next, how to phrase it; or they worry about what their supervisor might be thinking or whether their client thinks they are being helpful. Ironically, this preoccupation generally interferes with their ability to give the client their total attention—a prerequisite to being helpful. Counselors with excessive performance anxieties are less effective because they are responding more to their own internal need (to be helpful, competent, or liked) than they are to the client's need to be understood.

THE COUNSELOR AS A ROLE MODEL OF HEALTH

Observation and role modeling are among the most powerful processes through which human beings learn. We often learn more from watching someone's behavior than from what the person says. This is why demonstrations are such powerful learning techniques.

In Dan Millman's *Way of the Peaceful Warrior* (1984, p. 185), he relates the following story:

> A mother brought her young son to Mahatma Gandhi. She begged, "Please, Mahatma. Tell my son to stop eating sugar." Gandhi paused, then said, "Bring your son back in two weeks." Puzzled, the woman thanked him and said that she would do as he had asked.
>
> Two weeks later, she returned with her son. Gandhi looked the youngster in the eye and said, "Stop eating sugar."
>
> Grateful but bewildered, the woman asked, "Why did you tell me to bring him back in two weeks? You could have told him the same thing then."
>
> Gandhi replied, "Two weeks ago, I was eating sugar."

As a counselor, you are in a position to be a powerful role model. In facilitating clients' growth toward a positive self-concept, one of your most powerful tools is the role modeling you do: demonstrating your own positive self-concept and demonstrating the beliefs and behaviors you are attempting to help your client develop. If you cannot demonstrate yourself what you are attempting to help your client develop, there is no reason for either of you to believe you can be of help, as shown in the following example.

One of my student counselors had been working with a client who was trying to learn how to handle anger more positively. The client had become frustrated with her perceived lack of progress and had suddenly felt angry with the student counselor:

> *Client:* I don't think this counseling is working. (angrily) I think you should be able to help me more.
>
> *Student Counselor:* (defensively) Well, *I've* been doing everything I can; maybe you need to try a little harder.
>
> *Client:* (sudden realization) Wow. *You* don't deal with anger any better than I do. How can you help me?

SUMMARY

An effective helper is one who is able to facilitate human growth. The same principles of growth apply to both the helper and the one helped. Consequently, if helpers are unable to facilitate their own growth, it is less likely they will be able to help others.

The importance of helpers' personal growth to their ability to be effective can be summarized with the following:

1. *Parallel process of growth facilitation.* If the helpers cannot facilitate their own growth, how can they facilitate others'?

2. *Boundary distinction.* To help others, helpers must be able to distinguish their own issues from their clients' issues.

3. *Role modeling.* Role modeling is one of the most effective learning processes. The helper is in a powerful position for role modeling—whether it be good or poor psychological health.

Attitudes speak louder than words. Words cannot substitute for attitude. Attitude, therefore, is the most powerful message we communicate. Verbal skills are useless without effective attitude. Attitude reflects underlying beliefs. Beliefs and attitude, consequently, are the prerequisite foundation for effective communication.

One recalls the story of the Persian who took his son to a lovely garden where many people had gathered to pray. After an hour of chanting prayers, the boy looked around and observed that many of the worshipers were lost not in prayer but in slumber. He turned to his father and asked, "Are we not better than those who are sleeping instead of praying?" The father simply replied, "You might have been better had you not asked this question."

━━ Baha u Llah (Robert L. Gulick, Jr.),
The Seven Valleys and the Four Valleys

A Stage Structure for the Helping Process: Pre-Stage Attitudes of Rights, Respect, and Responsibility

A s I waited at the airport to board my plane en route to a convention, I overheard two students going home for spring break.

"But you can't do that!" I heard one exclaim. "Why not?" the second asked. "Well, because . . ." the first searched for a reason, "you wouldn't be able to finish your coursework." "I'll take an incomplete." "But then you wouldn't graduate this spring!" "So?"

The conversation had the unmistakable tone that comes when one individual believes and communicates that he or she knows what's best for another. It also reflected the predictable defensive response.

Our underlying beliefs about human nature, personal growth, and development have two major impacts: first, they affect how we view ourselves and approach our own growth; second, they become the foundation for how we perceive and respond to others. Our *beliefs* about human development and personal growth—our own and others—affect our *attitude* toward others and our *behavior* when responding and working with them. In the example above, for instance, the first student obviously held a belief related to finishing school, which was being imposed on the second student. It also was affecting how they related to each other. Many believe that one's beliefs are always transmitted to others, either directly or indirectly (Okun, 1992), and that one's beliefs and attitudes are the foundation of communication skills (Nelson-Jones, 1993). To the extent that our beliefs affect our view of ourselves, our attitude toward others, and our behavior when working with others, it is imperative to consider and evaluate our belief system.

8

Underlying Beliefs
and Personal Growth:
Rights, Self-Respect,
and Self-Responsibility

*[People] stumble over the truth from time to
time, but most pick themselves up and hurry
off as if nothing happened.*

— SIR WINSTON CHURCHILL

I n Part I of this book, the purpose, goals, and definitions of a helping relationship were addressed. In Part II, a system of outcome
and process goals for client, helper, and helping relationship was
presented, including goals in the affective experience, behavioral, and
cognitive dimensions. In Parts III, IV, and V, a stage structure for the
helping process is presented as a guide for facilitating the outcome goals
identified in Part II. Part III addresses underlying beliefs and attitudes
that lay the foundation for the communication model and helping skills
presented in Parts IV and V.

Beliefs are those tenets one accepts as true, and values are those
concepts one prizes, regards highly, or prefers (Cormier & Cormier,
1991). Our beliefs underlie our values, and our values are our attitudes
and feelings about what is good or bad (Okun, 1992). Our values determine how we act or don't act (Cormier & Cormier, 1991).

We all have beliefs and values. We have beliefs about human nature, development, growth, and goodness. It is impossible to be value free, and because we have beliefs, they go with us into the helping relationship. It *is* possible to make conscious decisions about the effect and imposition of our values and beliefs on others. Indeed, *unless* we are conscious of our beliefs and make conscious decisions about them, they *will be* transmitted, either consciously or unconsciously, and affect our client—either directly or indirectly.

A husband and wife go to see a counselor for marriage counseling. The male counselor asks the husband what he does for a living. He fails to ask the wife. The failure to ask reflects an assumption or belief, which the counselor holds, that wives don't work. The counselor didn't think to ask because he wasn't even conscious of the belief. If he were conscious of it, he'd realize it may not be shared by others, nor be applicable to these clients.

〰

What we do is based on a decision; what we decide is based on a belief; therefore, our beliefs are the foundation for our attitudes and actions.

〰

A STRUCTURE OF UNDERLYING BELIEFS FOR A POSITIVE SELF-CONCEPT

This structure operates on the assumption that our underlying beliefs affect our thoughts, behavior, and affective experience, and, therefore, our self-concept. In addition, it is predicated on the assumption that our underlying beliefs affect our attitude toward others, and, therefore, our work in helping relationships.

In the system of outcome goals presented in Part II, the overall outcome goal for clients was defined as a positive self-concept. The affective experience goals reflecting components of a positive self-concept were identified as self-acceptance, self-esteem, and self-actualization. The underlying behavioral goals reflecting these affective experience goals were defined as congruence, competence, and internal control. Beliefs underlying these behavioral goals were identified as rights, self-respect, and self-responsibility.

These underlying beliefs and attitudes, which are related to one's perspective on individuals' rights, respect, and appropriate responsibility-taking, are central to this structure and provide the foundation for the helping process. In this chapter, a conceptual framework of these underlying beliefs is presented, as it relates to individuals' perspectives on their right to be themselves, on having self-respect, and on assuming self-responsibility. These underlying beliefs are fundamental to a positive self-concept and they provide an underlying attitude central to an effective helping relationship. The three conceptual beliefs regarding rights, respect, and responsibility, presented in this chapter, are proposed as a structure of underlying beliefs for a positive self-concept.

Rights

A primary goal of most theoretical counseling orientations is self-acceptance or being true to oneself. Self-acceptance, however, is influenced by your basic belief about your *right* to be yourself. Take the classic example of someone trying to pressure another into doing something that person doesn't want to do:

> *John:* Hey, Sarah! I need to borrow your car. Mine's in the shop and I've got to run an important errand after work.
>
> *Sarah:* Well, John, I'd like to help you out, but I'm afraid I can't. (Sarah just doesn't feel comfortable lending her car.)
>
> *John:* Oh, no! Why not? I really need it, and I don't have anyone else who can help.
>
> *Sarah:* Well . . . (searching for an excuse to use to get her off the hook, because "just not feeling comfortable" seems unacceptable). Oh, I can't . . . because my dad won't let me lend my car (she defers responsibility to someone else).
>
> *John:* Oh, don't worry; he won't even have to know—I'll have it back in 30 minutes.
>
> *Sarah:* (feeling trapped) Oh, well, okay, but be careful with it.

Sarah did not feel she had the right to say "no" just because she wasn't comfortable letting John borrow her car. Consequently, she was not true to herself. She did not accept her feelings and did not act consistently with how she felt and what she wanted. Sarah's belief regarding her *rights* influenced her ability to be *congruent*, which reflects a lack of *self-acceptance*.

Right to Being

Conceptual Belief 1:
Each of us is inherently worthy and has the right to be and become who we inherently are, as long as we don't infringe on the rights of others in the process.

Right means "in accordance with justice, law, or morality . . . that which a person has a just claim to; power, privilege . . . that belongs to a person by law, nature, or tradition" (*Webster's New Collegiate Dictionary,* 1981). It is a democratic ideal that all persons have the right to pursue their own lives and destinies. The United Nation's Declaration of Human Rights in 1948 states, "All human beings are born free and equal in dignity and rights." America's Declaration of Independence, July 4, 1776, states that all "are created equal, that they are endowed by their creator with certain unalienable Rights, that among these are Life, Liberty and the Pursuit of Happiness."

The right for individuals to be who they are, and to live their own lives in a way in which they can feel comfortable, satisfied, and fulfilled, implies an acceptance of uniqueness and differences. It implies that individuals have the right to their own cultural, ethnic, spiritual, racial, gender-role, individual, and familial perspectives.

Few would argue with the basic tenet that as long as people do not violate the rights of others, all have the right to be true to themselves and to the pursuit of their own lives and happiness. The importance of this fundamental right of all individuals to have their own feelings, behaviors, and thoughts has been put forth as individuals' right to their own destinies, ideas, perspectives, lives, and interests (George & Cristiani, 1990); as an acknowledgment of individuals' right to their own uniqueness (Arkoff, 1988); and as an imperative for individuals to allow themselves to be themselves as a prerequisite to growing (Thompson & Stroud, 1984).

Many people struggle, however, with the application of this belief through congruence and self-acceptance. The extent to which we believe we have the right to be who we are affects our ability to be congruent, which in turn reflects the degree to which we can be self-accepting.

Respect

A primary outcome goal of most counseling orientations is self-esteem. Self-esteem is based on a basic respect for oneself—an experiencing of oneself as competent, capable, and able. It is based on self-approval (internal approval). There is no substitute for self-approval because no one else can give it. We often desperately seek a substitution through approval from others (external approval). As an example, in our cafeteria lunch line the other day, I overheard the following typical struggle:

Person A: I don't know; I *think* I handled it all right—the guy was just being a jerk, that's all. What else could I do? I don't think anyone could have dealt with an irate customer any better, do you? I mean, the guy was out of control. There really wasn't anything else I could do but call the manager.

Person B: I think you did the right thing. Of course no one else could have done any better. You did great.

Person A: But maybe I could have responded in a way that wouldn't have fueled his anger. If I had stayed calmer it probably would have gone smoother, and maybe I wouldn't have had to call the manager to bail me out.

Person A obviously does not feel she handled this situation as well as she could have. Approval and support from someone else, although comforting and perhaps helpful for other reasons, is *not* a substitute for the *self*-approval this person wants. She wants to be able to respect *her* competence, which results in *self-esteem.*

Respect for Capability and Worth

Conceptual Belief 2:
Each of us is capable, able, and inherently worthy of respect as a human being.

Respect means "to feel or show honor or esteem for; to hold in . . . regard; to consider or treat with deference or . . . regard" (*Webster's New Collegiate Dictionary,* 1981). Respect has historically been considered by most social and psychological scientists to be one of our deepest human needs (Fromm, 1982; Maslow, 1954; May, 1989).

Respect emerges out of Conceptual Belief 1—that all individuals have the right to be themselves. It reflects an acceptance of the right to one's own life and acceptance as a worthwhile person. It is the acceptance, despite differences, of all individuals' right to be themselves and the belief in individuals' abilities to make appropriate choices for themselves (Egan, 1990).

Inherent in respect is an acknowledgment and honoring of individuals' cultural, ethnic, spiritual, racial, gender-role, individual, and familial perspectives. Respect reflects an acceptance of the right of individuals to be themselves and live their own lives. Their choice of cultural, ethnic, spiritual, racial, gender-role, and familial perspectives is inherently part of this right.

Respect is nonjudgmental in nature. Respect reflects a genuine caring, free of judgmental evaluations of thoughts, feelings, and behaviors. It means focusing on objective understanding, without imposing value judgments or classifying feelings, values, or behaviors as good or bad (Patterson & Eisenberg, 1983).

Respect includes warmth, esteem, and positive regard. Some have used the term *warmth* to refer to the natural part of a respectful relationship (Egan, 1990); others have used *positive regard* when discussing the honoring of unique differences and the valuing of individuals' perspectives (Cormier & Cormier, 1991; Hackney & Cormier, 1988; Ivey, 1994). *To hold in regard or esteem* has also been used to refer to maintaining a positive view of the person, without being critical of limitations or differences (Patterson & Eisenberg, 1983).

The extent to which we believe we are capable, able, and worthy affects our ability to develop competence. Our ability to develop competence affects our self-esteem.

Responsibility

A primary goal of most theoretical counseling orientations is freedom, choice, self-realization, self-actualization, and self-responsibility. *Self-actualization* literally means actualizing or becoming oneself—that person one is innately. Actualizing oneself necessitates *choosing* to be oneself—making choices and decisions that allow one to be and do what one *is*. To make such choices, a person must possess an internalized sense of *control*. Internal control is an ability to *choose*, based on what feels right, resulting in self-approval, as opposed to seeking the approval of others. Internal control is based on the belief that only you are able to recognize and actualize yourself, and therefore, only you can logically be *responsible* for such actualization.

Consider the following typical example:

Mary: I don't know whether I should marry Bob and quit school to go with him like he wants, or stay and finish my degree.

Olivia: What do you *want* to do?

Mary: I don't know. I'm so swayed by Bob's rationale. He wants me to move with him now to his new job location and have a family. He says he'll make plenty of money at his new job to support us. It does sound tempting.

Olivia: But is that what you *want* to do?

Mary: Not really. I want to finish my degree first. I don't want to be dependent on Bob.

Olivia: So it sounds like you *know* what you want.

Mary: Yeah, I guess I do. It's just that Bob will be so upset—he has his heart set on my going with him.

Mary is struggling with a decision. She can choose to make a decision based on what feels right to her (internal control) or defer to what someone else (Bob) wants in order to please him (external control). Whichever choice Mary makes, she alone must accept responsibility for her decision, because she alone knows whether it reflects her true self and whether she can be happy with it. Mary's *self-actualization* is dependent on her ability to exercise *internal control* in making her decisions. Internal control reflects the belief that only she can be *responsible* for her decisions, since only she knows what would be congruent with her real beliefs.

Responsibility for Choices

Conceptual Belief 3:
Each of us is responsible for our own choices and for making decisions regarding who we are and how we live our lives.

Responsibility means "a condition . . . or instance of being responsible . . . answerable, or accountable; involving obligation or duties" (*Webster's New Collegiate Dictionary,* 1981). *Responsible* is "that which can be charged with being the cause, agent or source of something." Literally, responsible means "response able," "able to respond," or the "ability to choose a response." Self-responsibility, then, means that you are not only *able* to respond for yourself, but that you are *answerable* and *accountable* for yourself. In other words, if you are not being true to yourself, not making choices to actualize yourself, or not living your life in a way that you are happy with, then *you* are the one accountable.

Self-responsibility is built on Conceptual Beliefs 1 and 2, that you have a right to be yourself and that you are capable and able to make choices to actualize yourself. Self-responsibility is at the core of growth and change. The saying, "You can lead a horse to water, but you can't make it drink," acknowledges that the horse has control over whether he drinks, because only the horse can *do* it. Consequently, whether the horse drinks is the horse's responsibility. For an individual to grow and change, the *individual must choose to do the changing.* No one else can do it for another individual. Consequently, the individual alone carries the responsibility for choosing to change, and *whatever* the resulting actions are, that person is responsible for them if he or she is being responsible (Løvlie, 1982).

> ❧
>
> Change is a matter of choice and choice is a matter of responsibility.
>
>   ❧

Being responsible means exercising internal control in decision making. Only individuals themselves can know whether they are being true to themselves, happy with who they are, and happy with how they are living their lives. Consequently, only individuals themselves can recognize and decide what they need to be and do to be true to themselves. Ownership and responsibility for personal decisions and goals can rest *only* with the individual (Hutchins & Cole, 1992).

We learn to focus externally for approval during our childhood. Therefore, it feels natural to defer to others for approval and for decision making in adulthood. To the extent that we defer to others in decision making, we defer responsibility to them as well.

A natural consequence of deferring to others is to expect them to take responsibility for us, wait for them to take care of us, and then blame them when the outcome is not to our liking. The "quick-to-sue" trend in American society is reflective of the belief that someone else is responsible for our well-being, and therefore, someone else is to blame and accountable if something happens to us. The following is an example:

> A recent lawsuit against a major national park involved a man who had been drinking while walking down one of the trails into the park, tripping over a rock and falling off the trail. He held the park responsible for the rock's being on the trail and therefore responsible for his tripping and falling.

This example is indicative of the belief that people don't have to be responsible for themselves—for using good judgment regarding when to drink, or for watching where they walk—but can defer that responsibility to others. With the deferral of responsibility comes the deferral of accountability.

Responsibility means not blaming the past or others. As children we initially *do* depend on and defer to others for our well-being. Consequently, it's easy either to (1) want to continue to defer the responsibility for our happiness to others, and/or (2) blame the past for our present difficulties. John-Roger and McWilliams (1991) present an appropriate analogy to blaming the past for our struggles in the present:

> Imagine yourself dropping a glass and watching it break. Blaming gravity for breaking the glass is like blaming the past for your life today. It's true that without gravity the glass wouldn't have fallen, but you *know* about gravity and know about glasses.
>
> Childhood and your past are like gravity—they were what they were. Your life today is like the glass. If you drop it and it breaks, it's not gravity's fault. (p. 333)

Responsibility means not blaming circumstances, conditions, or conditioning for behavior. It means believing that behavior is the product of conscious choice, based on values, rather than a product of conditions (Covey, 1990). Being responsible means believing *we* must take

charge of being and doing what we believe we need to be and do, in order to feel satisfied with ourselves and our lives.

Responsibility means being accountable for our decisions, self, and happiness. The responsibility for individuals' happiness remains theirs, even if someone else is willing to accept responsibility for them (Warton & Goodnow, 1991). Just as no one can drink for the horse, no one can be happy for you—even though someone might be willing to try.

We are accountable for our choices. Our choices create who we are and how we live our lives—and ultimately determine our happiness. Responsibility means accepting that we have control over what we choose and how we process and perceive ourselves and our lives. Responsibility means accepting that whatever happens to me in my life, I had *something* to do with it: I either "created, promoted or allowed it" (John-Roger & McWilliams, 1991, p. 127).

Responsibility means taking charge of our well-being. When individuals are responsible, they take care of their well-being: emotionally, intellectually, physically, socially, and spiritually. When individuals are responsible for themselves, they do not do anything to injure, degrade, humiliate, or destroy themselves or another, and will not hold another responsible for their actions (Satir, 1988). The extent to which we believe we are responsible for our own choices and decision making affects our ability to develop internal control—which in turn affects our ability to actualize ourselves.

SUMMARY

Underlying beliefs influence our perspective on who we are (self-concept) and on our personal growth. The structure of underlying beliefs for promotion of a positive self-concept presented here include the following:

1. *Right to Being.* Each of us is inherently worthy and has the right to be and become who we inherently are, as long as we don't infringe on the rights of others in the process.

2. *Respect for Capability and Worth.* Each of us is capable, able, and inherently worthy of respect as a human being.

3. *Responsibility for Choices.* Each of us is responsible for our own choices and for making decisions regarding who we are and how we live our lives.

Underlying beliefs regarding human beings affect not only our perspective on self (self-concept) but also influence our attitudes toward others (i.e., clients). The translation of beliefs into attitudes is addressed in the next chapter.

> *The last of the human freedoms—to choose one's attitude in any given set of circumstances, to choose one's own way.*
>
> ━ VICTOR FRANKL

9

Attitudinal Goals in Facilitating Personal Growth: Rights, Respect, and Responsibility

It's not what we say, but the attitude that lies behind what we say.

— MARIANNE WILLIAMSON

O ur underlying beliefs about rights, respect, and responsibility not only affect our self-perceptions and personal growth, as discussed in the previous chapter, but they also affect our perception of others. Because our underlying beliefs are about humans, they affect how we feel about ourselves and others, as we are all human. When we work with others, our beliefs are reflected in our attitude toward them. For example, if I don't believe *I* have the right to feel or express my anger, that will influence my attitude toward *you* if and when you express your anger.

ATTITUDE AS PRIMARY COMMUNICATOR

What we think and feel is communicated powerfully to others in many more ways than by our words alone. Our attitude toward something or someone is demonstrated overtly and subtly, often without our knowledge. Because of the potency of attitude, we need to understand thoroughly what it is and how it operates.

109

Attitude Defined

Attitude refers to individuals' affective perspective or disposition, reflecting their general expectancy of a situation, based on their underlying beliefs. If they believe that the constructive expression of anger is healthy and appropriate, their attitude will likely be accepting and positive when anger is constructively expressed. If they believe anger should not be expressed, their attitude is likely to reflect unacceptance, fear, judgment, criticism, or horror when it is expressed. Their attitude reflects their beliefs about how they think things should be.

What we give to ourselves, we give to others.

— WILLIAMSON, 1992

Underlying Beliefs Affect Attitudes toward Others

If people believe they have the right to their beliefs and values, their attitude logically would be that others have a right to theirs. The underlying belief is that *humans* have a right to be who they are, and therefore, to their own respective beliefs and values. If, on the other hand, they do not believe they have a right to their own beliefs and values but must defer to some external source, such as parental or societal expectations, then their underlying belief is that *humans* must adhere to externally prescribed values. The belief applies not only to themselves but becomes their attitude toward others.

Attitudes, which are primarily nonverbal, generally come across regardless of one's verbal statements or behavior. If adolescents comply in returning the car keys, but slam them down with an air of defiance, they send a strong message through attitude. The attitude reflects a belief, perhaps, that they have a right to the car and that they are being unfairly treated.

Attitude versus Skills

Attitude is the component of communication that is an affective perspective based on cognitive beliefs; it is difficult to measure or observe directly. *Skills*, on the other hand, are behavioral in nature and therefore observable. Some skills referred to in the counseling literature, such as respect, unconditional acceptance, positive regard, and attending, are really more attitude than behavior. Attitude is primarily affective perspective based on underlying beliefs, with behavioral aspects resulting secondarily.

ૐ

Skills are based on behavior. Behavior is influenced by attitudes; therefore, skills are influenced by attitudes.

ૐ

Attitudes Are the Foundation for Skills

Underlying beliefs affect how people view themselves; how they view themselves influences their attitudes toward others; and attitudes affect or override behavior, and therefore, skills. In this framework, attitudes are considered the foundation for effective helping. Attitudes and skills are parallel to nonverbal and verbal communication. Skills and verbal content are concrete, specific, and direct. Attitudes and nonverbal communication tend to be vague, indirect, and difficult to interpret. A friend says to you, "Thanks a lot for asking my ex-boyfriend to your party." The content is a clear message, but depending on emphasis, intonation, facial expression, and body language, the true meaning could be the opposite of the content. The nonverbal communication will override the verbal content. Similarly, attitudes can override skills. Verbal communication is less effective without the nonverbal communication as the context. Just as nonverbal communication is the foundation for verbal content, attitude is the foundation for skills. *It is imperative, therefore, to be aware of your attitude, because your attitude is the primary message you communicate.*

RIGHTS, RESPECT, AND RESPONSIBILITY: FOUNDATIONAL ATTITUDES FOR HELPING SKILLS

This structure operates on the assumption that underlying beliefs affect self-concept and subsequent attitudes toward others. It proposes that our underlying beliefs regarding (1) the right to being, (2) respect for capability and worth, and (3) responsibility for choices apply not only to ourselves but are also translated into attitudes toward others, including our clients.

In the conceptual framework of outcome goals presented in Part II, the overall outcome goal for helpers was defined as being an effective helper. The affective experience goals reflecting components of an effective helper were identified as a role model, catalyst, and facilitator. The underlying behavioral goals reflecting these affective experience goals were defined as genuineness, positive regard, and focusing through empathy and expression. The attitudes underlying these behavioral goals were of rights, respect, and appropriate responsibility-taking (see Table 9.1).

Affective Experience Goals of an Effective Helper: Role Model, Catalyst, and Facilitator

An effective helper has been presented as incorporating three affective experience goals: to be a role model, catalyst, and facilitator. As a role model, the helper demonstrates and models an effective and healthy way of being, doing, and choosing. As a catalyst, the helper, by being in the relationship (i.e., providing positive regard), offers an ingredient in the helping relationship that is conducive to change in the client. As a facilitator, the helper, by conscious and intentional interventions and behavior, actively promotes client change.

Behaviors Underlying Affective Experience Goals: Genuineness, Positive Regard, and Focusing

There are behaviors associated with the outcome goals of being a role model, catalyst, and facilitator that can be observed and affected.

Genuineness is role modeling. Being genuine is being congruent in the presence of another. Genuineness, therefore, is role modeling congruence for the client.

Positive regard is a catalyst. Communicating positive regard for clients provides a catalytic environment in which they can view themselves as capable and worthy. If someone else, the helper, treats them with regard, they are invited to treat themselves with regard.

TABLE 9.1   Outcome Goals Related to Rights, Respect, and Responsibility

	Attitudinal Goals	Rights	Respect	Responsibility
Client Growth Goals	Behavioral Outcome	Congruence	Competence	Internal Control
	Affective Outcome	Self-Acceptance	Self-Esteem	Self-Actualization
Helper Skill Goals	Behavioral Outcome	Genuineness	Positive Regard	Focusing: Empathy and Expression
	Affective Outcome	Role Model	Catalyst	Facilitator
Helping Relationship Goals	Behavioral Outcome	Rapport	Processing	Directionality
	Affective Outcome	Trust	Understanding	Change

Focusing through empathy and expression is facilitating. The communication process of focusing through use of empathy (clients' perspective) and/or expression (helper's perspective) (see Part V) is active and intentional facilitation of growth.

Attitudes Underlying Behaviors and Affective Experience: Rights, Respect, and Responsibility

Rights. In order to model genuineness, helpers must be congruent. They must believe in the right to be themselves. This right applies not only to themselves but extends to their attitude toward the client.

ಞ

A basic belief must be that we all have the right to our own beliefs.

ಞ

Conceptual Attitude 1: Right to Being

Conceptual Belief 1 applies to you: You have a right to be who you are. You have a right to your own beliefs.

Respect. Positive regard for the client reflects an attitude of respect on the part of the helper. It means the helper believes the client is inherently worthy of respect as a human being.

Conceptual Attitude 2: Respect for Capability and Worth

Conceptual Belief 2 applies to you: You are capable, able, and inherently worthy of respect as a human being.

Responsibility. Taking appropriate responsibility for helping the client to focus and grow, through purposeful selected responses of empathy and/or expression, without taking responsibility *for* the client by rescuing or blaming (see Chapter 12), reflects appropriate responsibility-taking.

Conceptual Attitude 3: Responsibility for Choices

Conceptual Belief 3 applies to you: You are ultimately responsible for your own choices and for making decisions regarding who you are and how you live your life.

Unless we are aware of our beliefs and attitudes we will unconsciously impose them on others. Three steps to becoming conscious include (1) examining your own underlying beliefs, (2) making a con-

scious decision regarding the translation of your beliefs into attitudes, and (3) paying attention to the consistency, or consonance, of your verbal and nonverbal communication. Your nonverbal communication reflects your beliefs and attitudes and overrides the verbal content of what you say. What message are you sending?

SUMMARY

Attitudes are a powerful form of communication and can counteract and override words and behavior. Attitudes, therefore, must be viewed as a foundation for verbal and behavioral skills.

The underlying beliefs regarding rights, respect, and responsibility, discussed in the previous chapter, influence attitudes in the helping relationship. The following conceptual attitudes are proposed as a prestage foundation of the framework for the helping process:

1. *Rights.* You (and your client) have a right to be who you are. You have a right to your own beliefs.

2. *Respect.* You are capable, able, and inherently worthy of respect as a human being.

3. *Responsibility.* You are ultimately responsible for your own choices and for making decisions regarding who you are and how you live your life.

We may agree with the importance and validity of these attitudes in theory, but often they are challenging to implement in practice. Rather than communicating to our clients an attitude of acknowledgment of their *rights,* we attempt to control. We say we *respect* their uniqueness, but we still find ourselves feeling judgmental. Instead of allowing them to take *responsibility* for their life choices, we find ourselves trying to rescue them. The following chapters address the challenge of implementing the attitude of *rights, respect,* and appropriate *responsibility.*

10

Rights versus Control

*The uncreative mind can spot wrong answers, but
it takes a creative mind to spot wrong questions.*

━━ ANTHONY JAY

C onceptual Belief 1 and Conceptual Attitude 1 state that individuals have the *right* to be and become who they inherently are, as long as they don't infringe on the rights of others. This means they have the right to

- Have the courage to be
- Be themselves
  Be true to themselves
- Congruency of thoughts, feelings, and actions
- Self-acceptance (valuing oneself regardless of perceived shortcomings)

As this belief applies to all humans, it applies to *both* client and helper. The right is to be oneself and to control one's *own* life, but not another's.

The antithesis of this belief is that individuals do *not* have the right to be themselves. It is that someone else knows best how they should live their lives, and consequently needs to manage or *control* their lives for them. In the helping relationship, there are rights specific to both the client and helper roles, and there is also a great deal of potential for the attempt to control.

CLIENT RIGHTS

Applied to clients, Attitude 1 means clients have the right to their own beliefs and values, to be and become who they are, and to live their lives as they choose. These rights, when viewed as specific to the helping relationship, are consistent with (1) the facilitation of personal growth, self-responsibility, and individual choice; and (2) ethical treatment, including competent counseling and the right to privacy.

Facilitation of Personal Growth, Self-Responsibility, and Individual Choice

Individuals each have a life like no other life (Arkoff, 1988). The right to be oneself means the right to become one's unique self. Ideally, this means that individuals have the right to develop fully their unique capabilities and potential through the ongoing process of self-actualization. Acknowledgment of this right means that helpers accept their clients' uniqueness, including cultural, ethnic, spiritual, and familial perspectives, ability to solve their own problems, and right to make their own choices.

Because all individuals are unique, and *in order to actualize* their true selves, they *must* have freedom of choice. The American democratic ideal ensures the right of all persons to control their own destinies, to pursue their own interests, to make their own choices. The right to choose freely and the right to veto decisions made for them is consistently viewed as a right of clients (Corey, Corey, & Callahan, 1993; George & Cristiani, 1990; Wright, 1987).

Clients have the right to services that have as the primary goal the clients' personal welfare, including the facilitation of their personal growth, self-responsibility, and individual choice (Van Houten, Axelrod, Bailey, & Favell, 1988). Clients have the right to have their perspectives respected by the helper, and to expect that the helping relationship will increase their ability to function more responsibly, effectively, and independently (Corey, Corey, & Callahan, 1993).

Ethical Treatment Included in Client Consumer Rights

Client consumer rights identified by the National Board for Certified Counselors (1989) and Chi Sigma Iota include (1) informed consent,

(2) freedom of choice and participation, (3) competent counseling based on professional ethical guidelines, and (4) the right to privacy.

Informed consent means that clients are made aware of relevant information that might affect their decision to agree to counseling. Relevant information includes

- qualifications, credentials, approach, techniques, and limitations of the counselor
- logistical information and expectations regarding billing, time schedule, cancellations, and emergency procedure
- limitations of confidentiality due to thoughts, actions, or intent to harm
- access to referrals, records, code of ethics, and appropriate professional organizations

Freedom of choice and participation means clients are vital members of the team working on their behalf. They make their own choices regarding participation—from goal setting to evaluating outcome and ultimately to terminating.

Clients' right to competent counseling means first and foremost "Do No Harm." It is essential that the client be able to trust the counselor, since trust is the basis for clients to divulge information in a way that makes effective helping possible. To offer a service that one is incompetent to deliver is a violation of client rights (Patterson & Eisenberg, 1983). Clients have the right to a therapeutic environment and effective helping procedures.

The right to privacy and confidentiality is perhaps the most basic right of clients: the guarantee that what they choose to reveal or not reveal within the helping relationship will be respected. Such respect takes two forms: acknowledging the clients' right to privacy (1) *within* the helping relationship and (2) *outside* the helping relationship. Privacy within the helping relationship means the clients decide what, when, and how much they are ready to divulge. Counselors who "coerce clients to reveal information [they] are not yet ready to reveal [are] guilty of an invasion of privacy" (Patterson & Eisenberg, 1983, p. 10). Respecting the right to privacy outside the helping relationship means maintaining a confidential relationship, allowing clients to trust that they can freely reveal significant information. It is a violation of privacy for a counselor to reveal to another person information that was given in confidence within the helping relationship, except under life-threatening circumstances (American Counseling Association, 1988).

Ethical practice can be defined as responsibly providing a helping service for which one has been appropriately trained and during which one adheres to ethical standards of care. Patterson and Eisenberg (1983, p. 10) identify unethical practice as occurring under three conditions: "When problems are beyond the scope of [the helpers'] training; when helpers exploit their position to collect fees or salary for incompetent service; or when helpers fail to understand their obligation to respect a client's right to privacy and to have free choice."

HELPER RIGHTS

Ethically and professionally, counselors have the responsibility to provide competent counseling, adhering to ethical guidelines, as discussed in Chapter 12. Helpers have the same basic rights, proposed in Conceptual Belief 1, as all individuals. Applied specifically to the helper role, this includes the right to (1) congruence, (2) ethical treatment, (3) appropriate responsibility-taking, and (4) non-negotiable items.

Congruence

The right to be oneself, applied to helpers, means helpers have the right to be congruent with their own thoughts, beliefs, feelings, values, actions, and behaviors. It does *not* mean the right to *impose* their beliefs on the client, but rather the right to be clear and congruent with themselves.

Ethical Treatment

The right to ethical treatment of the counselor means the counselor has the right to set

- guidelines regarding acceptance, termination, and referral of clients, based on background, qualifications, and ability to serve client needs
- limits regarding physical and verbal abusiveness
- appropriate ground rules regarding interactions with other clients and/or the counselor

- contractual agreements regarding fee payment, time commitments, cancellation policy, and concurrent therapy with other counselors
- boundaries regarding dual relationship with clients

Appropriate Responsibility-Taking

The helper can facilitate growth and change, but the client has to do the growing and changing. Appropriate responsibility-taking means that helpers take responsibility to do what is appropriate to their role—that is, establishing a therapeutic environment, providing competent counseling, adhering to ethical standards, and facilitating client growth; clients take responsibility to do what is appropriate to their roles.

Appropriate responsibility-taking for the client includes making the choice to enter into counseling—even in involuntary referrals, the client has a choice of entering counseling or some other alternative—participation in goal setting, engaging in the helping process, and growing and changing.

Helpers cannot force clients to change, nor is it their responsibility (or right) to attempt to do so. The helper has the right to expect appropriate responsibility-taking from the client.

Non-Negotiable Items

The helper needs to be able to recognize and say "no" to items that are non-negotiable. Such items include requests that are illegal, inappropriate, harmful to others, not possible, or unethical (Johnson, 1993). When in doubt, the helper has a right to postpone a response to think about the issue or to consult a colleague and/or a supervisor.

CONTROL

Few would argue, and most would agree in theory, with Conceptual Belief/Attitude 1: All individuals have the right to their own beliefs and values, and a right to actualize themselves, short of harming others. In practice, however, many experience difficulty in applying this principle.

Belief versus Behavior

One of my practicum students reported the following scenario during supervision:

> When my client said she was waiting for her husband to decide whether she could continue working, I was shocked. She should decide that. I couldn't believe she would willingly give her power away like that. My response reflected my shock, too, I think. My voice even cracked when I said, "You don't mean that you're letting your husband decide whether or not you work, do you? That's *your* decision and *you've* got to make it."

This exemplifies a situation in which counselors *say* their clients have the right to be themselves and make their own decisions, yet in practice the counselors impose *their* beliefs on the clients. The counselors obviously believe that their views are best for their clients and therefore are attempting to *control* their clients' perspectives and actions. The irony is, as in this example, that while counselors say their clients need to be making decisions for themselves, the counselors themselves are attempting to make these decisions. It is important to assess whether your behaviors are reflecting your beliefs.

Power Within versus Power Over

To the extent that we believe we have "power within" ourselves, there's no need to exercise "power over" another. If we truly believe we all have the right to be ourselves, we have no need to try to control others' rights to be *themselves*. Indeed, "duping clients into living more effectively is, from a values perspective, a contradiction in terms" (Egan, 1990, p. 65).

Process versus Outcome

> *Where we have an attachment to results we*
> *tend to have a hard time giving up control.*
>
> WILLIAMSON, 1992

The Conceptual Belief/Attitude that individuals have the right to be themselves means counselors can invest in facilitating the clients' *process* of self-exploration, but not the *outcome*. Only clients can recognize the outcome that is consistent with, and appropriate to, who they are; therefore, the helper must detach from the resulting outcome. To the extent that helpers think they know what the outcome "should" be—that is, how the clients should be—they are more likely to try to control the process—that is, get the client to achieve the outcome. Trying to control the outcome is a contradiction of Conceptual Belief/Attitude 1.

SUMMARY

The conceptual belief and attitude that individuals have the right to be who they are applies to both helper and client. Rights specific to clients include the right to the facilitation of personal growth, self-responsibility, individual choice, and ethical treatment. Rights specific to the helper include the right to congruence, ethical treatment, appropriate responsibility-taking, and non-negotiable items. The antithesis of the belief and attitude that individuals have the right to be themselves is control.

11

Respect versus Judgmentalness

Most of our assumptions have outlived their usefulness.

— MARSHALL McLUHAN

C onceptual Belief and Attitude 2 is that individuals are capable, able, and inherently worthy of *respect* as individuals. It follows that if individuals have the right to be who they are (Conceptual Belief 1), they need to have their choices respected and they are deserving of positive regard as unique individuals. Respect relates to one's

- Courage to act
- Doing and acting with internal approval
- Developing abilities, capability, and competency
- Self-esteem (valuing oneself based on perceived strengths)

Applied to the helper, this belief means self-respect. Translated into attitudes, it means respect for the client's inherent worth, uniqueness, and capability.

The antithesis of this belief is judgmentalness. Judgmentalness is the tendency to form critical opinions about another based on the imposition of *your* beliefs and values. Judgmentalness communicates implicitly that your belief or perspective is "better" than another's.

Culturally, respect is associated with power and economic terms of value such as *owe, give, demand,* or *deserve.* This is not surprising, as respect is one of our deepest needs. In the helping relationship, the basic tenets of respect are the same for both helper and client, although the function of respect takes different forms in each respective role.

CLIENT RESPECT

Applied to clients, Conceptual Attitude 2 means the acknowledgment that our clients are inherently unique and worthy individuals, with the capability and capacity to problem solve, to develop their abilities, and to do what they need to do to be who they are and to live their own lives. Applied to clients, respect includes (1) acknowledgment of their right to individual uniqueness, (2) positive regard, (3) suspension of judgment, and (4) acknowledgment of their individual capability for self-responsibility.

Acknowledgment of Right to Individual Difference and Uniqueness

Respect, in large part, is simply the acknowledgment of Conceptual Belief/Attitude 1: that people have the right to be themselves—unique individuals. It is openly, honestly acknowledging, appreciating, and tolerating differences (Ivey & Simek-Downing, 1980). It means helper awareness of personal values and an ability to accept and appreciate people with different values (Gazda, Asbury, Balzer, Childres, & Walters, 1984). It includes accepting and honoring unique cultural, ethnic, racial, spiritual, gender-role, and familial differences. It means prizing the individuality of clients and supporting their search for self (Egan, 1990).

Respect means the acceptance of clients' values and right to make choices based on them, even if these choices are different from the helper's. These decisions and choices include the clients' right to determine the goals of counseling.

Positive Regard for Inherent Worth

Respect means positive regard. It means acceptance and acknowledgment of the client as an inherently worthy fellow human being. It means "prizing" people simply because they are human (Egan, 1990), accepting people as "dignified, worthy individuals" (Gazda et al., 1984).

Respect necessitates the ability to differentiate between people's inherent worth as individuals and their values or behavior. Positive regard is often misconstrued as agreeing with or condoning a behavior when in fact it simply means accepting your clients as inherently worthy fellow human beings. It does not mean you have to accept their behaviors or concur with their values (Gazda et al., 1984); neither does respect mean encouraging or accepting behavior that is self-destructive or harmful to others.

Positive regard is not the same as taking the clients' side or acting as an advocate. Positive regard is a genuine caring about your clients. It means communicating a "deep and genuine caring" for them as people (Egan, 1990). It means not indifference but the offering of time and energy because their well-being as humans matters (Patterson & Eisenberg, 1983).

Positive regard is reflected in a sincere effort to understand your clients and their perspectives. "One of the best ways of showing respect is by working to understand clients—their experiences, their behavior, their feelings" (Egan, 1990, p. 67). Respect is entering the clients' world and trying to understand their experience. Clients should feel that the helper is trying to understand them and that they and their experience are being regarded with concern.

Suspension of Judgmentalness

Judgmentalness is the antithesis of respect, acceptance, and positive regard. Respect, therefore, requires the suspension of critical judgment.

> *You are there to help clients, not to*
> *judge them.*
>
>   ━ Egan, 1990

Everything the helper does should promote the process goal of client growth and self-understanding, and the outcome goal of client positive self-concept. Seldom does critical judgment promote these goals.

Respect means trying to understand your clients rather than to judge them.

Suspension of judgmentalness means not classifying client values as good or bad. It means trying to *understand* the thoughts, feelings, and behaviors of your clients without imposing value judgments. Effective helpers try to understand how their clients' behaviors might reflect their way of coping with some life circumstance. Even though helpers will develop opinions regarding the effectiveness of the behavior or decision, they must refrain from classifying it as good or bad (Patterson & Eisenberg, 1983).

> *Neither judge nor condone the client's behavior.*
>
> ⟵ EGAN, 1990

Acknowledgment of Individual Capability for Self-Responsibility

Respect implies an acknowledgment of the client's capability for self-responsibility. It is based on the assumption that individuals are not only capable but are the only ones able to take responsibility for their own lives and happiness. It means acknowledging clients' unique talents and abilities, and not devaluing them due to their limitations or differing perspectives.

Respect in the helping relationship also refers to a belief in clients' abilities and desire to grow, learn, and develop into responsible, self-reliant individuals. It often means placing, or helping the client place, demands on the client. As Egan (1990, p. 69) points out,

> Counselors show respect by helping clients through their pain, not by helping them find ways to avoid it. . . . Respect includes an assumption . . . that the client . . . is [desirous] of living more effectively, respect places a demand on the client . . . while at the same time offering help to fulfill the demand.

As a result of the belief in clients' capability for self-responsibility, the counselor must defer to the clients' agendas and trust their ultimate ability to know what they need.

> Effective helpers do not have hidden agendas or ulterior motives. (Patterson & Eisenberg, 1983 p. 13)

The only meaningful question is whether a concept is helpful to a client. (adapted from Williamson, 1992)

Keep the client's agenda in focus. (Egan, 1990, p. 61)

HELPER RESPECT

Respect essentially refers to the acknowledgment and acceptance of one's rights (as conceptualized in Belief/Attitude 1), one's capabilities and inherent worth as an individual. Consequently, helper respect refers first to the acknowledgment and acceptance of helper rights, as discussed previously. Perhaps more important, helper respect must begin with *self-respect*. Self-respect applied to helpers means they experience themselves as competent and capable, not only in living their personal lives effectively but also in their role as helpers.

Helper self-respect is imperative for two reasons: (1) self-respect is a prerequisite to an attitude of respect toward others, and (2) the helper is role modeling self-respect for the client.

JUDGMENTALNESS

One day during my class in communication skills, I had been demonstrating the attitude of respect. Following the demonstration, two students were doing a role play, and the student in the counselor role was attempting to implement a respectful attitude. The student playing her client had done something she strongly disagreed with and she was having an especially difficult time because she felt judgmental. Following the role play she burst out, "I can't help it, I'm always judging!"

We are, in fact, always judging. Once we make a judgment, however, the very next judgments need to be (1) What's my goal? and (2) How should I respond to facilitate that goal?

The student counselor above disagreed with her student-client's decision to quit his job and move back home with his 3-year-old son, letting his parents support him through school. She assessed the situation and made a judgment about it based on the information she was given. Such assessment is a natural and important part of being able to be helpful. It is what *follows* this judgment that determines whether one is

EVERYBODY GETS JUDGED ENOUGH WITHOUT A THERAPIST JOINING IN.

implementing an attitude of respect or judgmentalness. Often we need to be reminded of our goal. Our goal here is *not* to pass judgment on our client or to decide what *we* would do or think our client *should* do, but rather to help facilitate our client's growth through self-understanding and self-responsibility. How we facilitate growth, as discussed in Part IV and Part V, includes an attitude of respect for our clients. This means helping *them* decide what to do and respecting their decisions.

Judgment is an assessment of circumstances, situations, thoughts, feelings, behavior, and actions from the viewer's perspective. In the helping relationship, helpers are making assessments continuously. They are assessing consonance between what a client is saying and doing; they are assessing congruence between thoughts, expressed feelings, and actions; they are assessing consistency of verbal and nonverbal behaviors. They are assessing whether a client's statement indicates probable harm to self or others. They are making a judgment about what response to make to best facilitate growth. Judgments are an appropriate and integral part of the helping process.

Judgmentalness is a propensity toward or an excessive imposition of one's judgments on others. It implies a right or wrong, a good or bad. Judgmentalness focuses on the person rather than the circumstance. Judgmentalness implies a criticism, a put-down, or an attitude of "being better than" or "knowing more than" another. In a helping relationship, it implies that the counselor knows best how the client should think, feel, or behave; and therefore the helper is in the best position to determine what the client should think, feel, or do.

Respect versus Judgmentalness

Judgmentalness is the antithesis of respect. Respect means the acknowledgment of and positive regard for individuals' rights to be their unique selves, the suspension of critical judgment, and the acknowledgment of individuals' ability for self-responsibility. Judgmentalness imposes one individual's values, beliefs, and standards on another.

Distinction between person and behavior or circumstances is fundamental to respect. If a child spills a glass of milk or trips and falls, the behavior or circumstance must be distinguished from the inherent goodness of the child. The parent who shouts "You are a clumsy child" is making the behavior synonymous with the child. A response such as "I know you're trying to be careful; sometimes even when we're trying hard, we still have accidents" helps distinguish between the person and circumstance.

Professional judgment means assessing circumstances, behaviors, thoughts, feelings, events, and context, and making decisions regarding what behavior and approach to take, as a counselor, to best facilitate client growth. Professional judgment is not judgmental. Professional judgment works in tandem with respect for the client as a person, regardless of circumstance or behavior.

SUMMARY

The conceptual belief and attitude that individuals are capable, able, and worthy of respect as unique individuals applies to both helper and client. Respect specific to the client includes an acknowledgment of the right to individual difference and uniqueness, positive regard of inherent worth, suspension of judgmentalness, and acknowledgment of individual capability for self-responsibility. Specific to the helper, respect means acknowledgment and acceptance of one's rights, capabilities, and inherent worth, resulting in *self-respect*. Self-respect needs to be present both personally and professionally.

The antithesis of respect is judgmentalness.

12

Appropriate Responsibility versus Rescuing and Blaming

*Change is a matter of choice, and
choice is a matter of responsibility.*

━ VONDA LONG

C onceptual Belief and Attitude 3 is that individuals have *responsibility* for their own choices in life, in terms of who they are and how they live their lives. The rights of clients and helpers, as individuals, are the same: the right to be themselves and to live their lives in their own way (short of harm to others). Individual rights are centered around the issue of responsibility. If I have the right to be myself and to live my life in a way to be happy, and I want to exercise this right, then I must take responsibility for being myself and living in a way in which I am happy. The reason I *must* take responsibility for this right is because no one else *can* know who I am or what I need to be happy.

Again, the proverb "You can lead a horse to water, but you can't make it drink" remains an apt analogy. No one can drink *for* another, or know when that person is thirsty; no one can be happy *for* another, or know what that person needs to be happy. Individuals have the right to be and live their life choices; individuals are responsible for being and living their life choices.

Taking appropriate responsibility for oneself and one's life means

- Having the courage to choose
- Choosing to be and act in ways true to oneself

133

- Having internalized control and approval in making decisions
- Being self-actualizing—making choices that lead to increasing fulfillment and growth

Applied to the helper, this belief means self-responsibility and taking responsibility for one's role as helper without taking responsibility for clients and their choices. Translated into attitude, it is respecting clients' ability to take responsibility for their own lives and choices.

The antithesis of appropriate responsibility-taking is rescuing and/or blaming. It is denying individuals the opportunity to be responsible for themselves.

CLIENT RESPONSIBILITY

Self-responsibility and increasing clients' responsibility for their actions, thoughts, feelings, issues, happiness, and life is at the heart of the helping process.

> *It is assumed that clients are capable of making choices and to some degree controlling their [lives].*
>
> — EGAN, 1991

Self-responsibility is a core goal. All counseling orientations identify self-responsibility as a goal of the counseling process. According to Karen Horney (1945), the assumption of responsibility for oneself and one's life is one of the primary objectives of psychodynamic psychotherapy. The development of a sense of personal responsibility is viewed as a prerequisite to acknowledgment of one's responsibility to others (such as one's children), and as Adler (1963) emphasized, to society as a whole.

Humanism emphasizes the overall dignity and worth of human beings and their capacity for self-realization. Humanists believe that we contain within ourselves the potential for healthy and creative growth and that if, and only if, we are willing to accept responsibility for our own lives, we can realize this potential (Hall & Lindzey, 1985). A primary goal of the therapist is to motivate the client to assume responsibility for finding meaning in life (Burke, 1989).

Self-responsibility from a behavioral perspective means that clients ultimately take responsibility for their behavior and lives by learning how to identify and change maladaptive behaviors and/or environments. While others may influence individuals with rewards or punishments, ultimately the behavioral change, practice, and maintenance of a new behavior pattern rests with the individual.

Cognitivists view self-responsibility as clients taking responsibility for their lives by learning how to recognize and change thought patterns (content and/or process) that lead to undesirable feelings and/or actions. Cognitivists contend that taking responsibility to change maladaptive thought patterns and irrational beliefs is at the core of a healthy growth process (Ellis & Harper, 1975).

Self-responsibility is an existential reality. Because humans are capable of thinking, feeling, and problem solving, they must answer for, or be accountable for, their actions. The individual alone is ultimately and existentially responsible for his or her choices, life, and happiness.

> Responsibility and all that it entails is one of humankind's fundamental existential concerns. It is intimately tied to issues of personal freedom and meaning in life. Acknowledging one's freedom to construe life's meaning as one chooses carries with it a simultaneous and potentially overwhelming recognition of personal responsibility. (Yalom, 1980, p. 85)

The client alone can change. Just as the horse alone can drink, only the client can change. The counselor cannot change *for* the client, nor force the client to change. The client must choose to change. Consequently, the client is responsible for change.

> *Clients alone can choose self-responsibility*
> *for personal change. The client must assume*
> *ownership of and responsibility for that*
> *decision.*
>
> ⟵ HUTCHINS & COLE, 1992

The client has responsibility in the helping relationship. Clients' responsibility is to be accountable for themselves—their thoughts, feelings, behaviors, life, and happiness. This applies within as well as outside the helping relationship. Consequently, clients need to provide information to the counselor regarding whether and to what degree the helping process is meeting their needs and goals.

Clients also have ethical pragmatic responsibilities within the counseling relationship. These have been identified by the National Board for Certified Counselors and Chi Sigma Iota as follows:

- Setting and keeping appointments with the counselor. Letting the counselor know as soon as possible if you cannot keep an appointment.
- Paying your fees in accordance with the schedule you preestablish with the counselor.
- Following through with agreed-on goals.
- Keeping the counselor informed of your progress toward meeting your goals.
- Terminating your counseling relationship before entering into arrangements with another counselor (National Board for Certified Counselors, n.d.).

HELPER RESPONSIBILITY

The helping process involves shared responsibility. Some responsibility, such as ultimate change and growth, rests appropriately with the client. Other responsibilities are appropriately the counselors'. Counselor responsibility can be generally viewed as fourfold: (1) self-responsibility for his or her own personal growth, (2) separation of personal issues from client issues, (3) responsibility for ethical treatment, and (4) active facilitation of client growth. Although each of these is separate and distinct, there is also overlap between them.

Counselors are responsible for their own growth. Self-responsibility is at the core of growth and change. The counselor must possess self-responsibility in order to help the client develop self-responsibility. This is for two reasons. First, the counselor is a role model and is demonstrating for the client a way of being and behaving. Second, it is difficult, if not impossible, to help someone else learn something that you have not yet developed.

કૃ

You can only help another grow as far as you, yourself, have grown.

કૃ

Personal issues must be separated from client issues. In order to guard against projection, manipulation, or countertransference, helpers have a responsibility to distinguish between their own values, problems, and issues and those of their clients. If, for example, the client is going through a divorce and the counselor has been divorced, it is possible that the counselor might assume the client's experience is going to be similar to that of the counselor. As a result, the helper might project his or her feelings regarding the divorce process onto the client and make faulty assumptions regarding the client's experience and needs.

Counselors must ensure ethical treatment. Responsibility *is* an ethical issue (Brammer, 1988). Responsibility for ethical treatment and the protection of client rights is a major responsibility of the helper. The client, of course, shares the responsibility for the helping process and outcomes, but it is the helper's responsibility to ensure that the process adheres to ethical principles.

It is the counselor's responsibility not only to protect clients' rights but to provide information to clients about their rights. This includes informing them of significant facts, the nature of procedures, and probable outcomes and potential difficulties (Corey, Corey, & Callahan, 1993). Information regarding ethical guidelines, including confidentiality, limits of confidentiality, and duty to report, should be included as part of the clients' informed consent to participating in the counseling process.

&

The integrity and welfare of the client is central to ethical treatment.

&

Counselors must facilitate client growth. This is the primary responsibility of the counselor. How this is done, in terms of process and techniques, will vary depending on the counselor's theoretical counseling orientation.

Although the client's responsibility is to grow, the counselor's role includes the responsibility to facilitate that growth. It is not the counselor's responsibility to do the growing or coerce the client to grow. It is not the counselor's place to decide *for* the client what the client needs, or what the client should choose to do or decide. It *is* the counselor's responsibility to facilitate the client's decision making.

In general, facilitation of growth incorporates (1) the creation of a safe environment conducive to personal growth (including rights, respect, and appropriate responsibility-taking), (2) the development of a supportive, positive, egalitarian relationship with the client, (3) the identification of realistic and achievable goals for the client, and (4) the application of specific processes and skills for the purpose of facilitating client growth.

RESCUING AND BLAMING

Rights, respect, and responsibility go hand in hand. We have the *right* to be ourselves and to have *respect* for ourselves; therefore, the *responsibility* for our growth must logically rest with each of us. We may have many influences on us, but only we can be responsible for ourselves. That responsibility includes all our human dimensions: feelings, behaviors, and thoughts.

We are responsible for our own choices and for making decisions regarding who we are and how we live our lives (Conceptual Belief 3). Many, however, find this application quite challenging. In theory, we accept the right to be who we are, respect our worth and capability to make decisions, and take responsibility for both implementing our choices and accepting the consequences of them. In practice, however, there is a tendency to *rescue* others rather than let them take responsibility for themselves, and to *blame* others, rather than accept responsibility for our own decisions.

Rescuing

Rescuing interferes with the development of self-responsibility. True rescuing is an emergency measure used to help someone, typically in a life-threatening situation, who is unable to help himself or herself. It is meant as an emergency, *temporary* measure, when people are *unable to help* themselves. For our purposes, *rescuing* is defined as the voluntary, unnecessary assumption of responsibility for another's thoughts, feelings, actions, or life decisions. To develop a pattern of voluntarily rescuing clients when they *are* able to help themselves and don't need rescuing can foster a false sense of responsibility on the part of the counselor, discounts the clients' abilities to help themselves, and fosters a dependency that interferes with their development of self-responsibility. The responsibility of the counselor is to help clients develop self-responsibility, not to become responsible for them.

Rescuing versus responsibility is actually an ethical issue. How much responsibility can counselors ethically assume for their clients' attitudes and behaviors (Brammer, 1988) before they are discounting their rights, disrespecting their abilities, or interfering with their development of self-responsibility? One experienced counselor writes of an early pattern of taking too much responsibility for clients:

> The harder I worked, the less my clients worked. The more responsibility I took, the less I helped them take responsibility. (Hines, 1988, p. 106)

The helper must take a position of helping clients to take responsibility for themselves.

> *You can give a man a fish and he'll eat today;*
> *you can teach a man to fish and he'll eat for*
> *the rest of his life.*

> ANCIENT PROVERB

Rescuing is focusing on the need of the helper rather than the client. Most people attracted to a helping profession such as counseling have a genuine desire to be helpful. The potential problem with this genuine desire is in the definition of *helpful.* In the proverb above, one might consider "giving a man a fish" helpful. Indeed, it is helpful, if at the moment the man is starving and too weak to learn to fish. However, to continue to give him fish when he is capable and able to learn to fish is fostering dependency. The key is what is identified as the goal of helpfulness. In a helping relationship, the goal is facilitation of client growth and self-responsibility.

Helpfulness that fosters dependency instead of self-responsibility is generally focusing on the need of the counselor in one of two ways: (1) the counselor's need to be needed is motivating him or her to foster dependency, and/or (2) it is easier to help by "doing it" for people than to help them learn for themselves. Either motivation can result in a relationship in which the client is dependent on the helper.

Rescuing discounts the client's capabilities. If one has respect for individuals' abilities and capacity for self-responsibility, one trusts their capability to make choices and decisions regarding their lives. Rescuing discounts that capability.

Clients have more resources for managing problems in living and developing opportunities then they, or most helpers, assume. (Egan, 1990, p. 73)

Helping is a team effort in which the clients learn to take responsibility for their own growth and lives. It is the counselor's responsibility to facilitate this growth, helping their clients to utilize their own resources to help themselves.

Blaming

I know, for myself, that if I'm feeling a desire to blame, I'm either (1) not taking responsibility for myself, or (2) am voluntarily taking responsibility for—trying to rescue—another, and am judging that person's lack of responsiveness.

Blaming reflects judgmentalness. If helpers believe their clients have the right to be who they are, respect their abilities, and believe they are responsible for their own choices, the helpers have no need to blame their clients for their decisions. Blame implies a judgment of fault, right or wrong. There is no right or wrong in decisions relative to personal exploration and growth.

Blaming can follow rescuing. When one rescues another, there's a tendency to expect gratitude, appreciation, and compliance. Then when those rescued don't live up to expectations, rescuers feel like blaming them—generally for botching the outcome of the perceived heroic rescue! Take the following example:

> An intern of mine was working at a shelter for survivors of domestic violence. She had taken on special feelings and hopes for one young mother of three. She made special efforts to help set up appointments for job interviews, arrange transportation, and do everything she could to get this woman out of the abusive relationship and into a new and better life. She felt special pride when the woman, after much effort, was offered a job. She then felt a shocking defeat when the woman quit the job and went back to her abusive husband. The intern threw her hands in the air in utter disgust.

Here's a case where the counselor took on responsibility for the actions of her client, and when the actions did not meet her expectations, she blamed her client and was judgmental of her decisions.

SUMMARY

The conceptual belief and attitude that individuals are responsible for their own decisions and life choices apply to both helper and client. Appropriate responsibility, specific to the client, includes self-responsibility as a core goal and existential reality, the acknowledgment that the client alone can change, and appropriate responsibility-taking within the helping relationship. Specific to the helper, appropriate responsibility-taking includes self-responsibility, the ability to separate personal issues from client issues, responsibility for ethical treatment, and responsibility for the facilitation of client growth. The antithesis of appropriate responsibility-taking is rescuing and/or blaming.

> *It is no use to blame the looking glass if your face is awry.*
>
> ◄— NIKOLAI GOGOL

Communication skills are the core of the helping process. As patterns of communication have already been developed, the question is how they can be used consistently and effectively in facilitating the personal growth of the client.

ON LIFE

Golden sun
And salty breeze
Softly light
The rhythmic sea.

Through troughs and swells
I gaily rode
As sea heaved sighs
And currents flowed.

Then caught by a crest
And toppled to shore
I sat stunned and crumpled
On the sandy floor.

But golden sun
And salty breeze
Touch the sea
And beckon me.

A Stage Structure for the Helping Process: Communication Components and Model

C ommunication is complex. It involves two or more people with their respective perspectives, representing their values, beliefs, assumptions, needs, cultural, spiritual, and family backgrounds, expectations, interpretations, experiences, and both past and present thoughts, feelings, and behaviors. It involves one person sending a message to another while both are being influenced by their own perspectives.

Even a response to a simple statement, like "That's a nice dress," requires the individual to process the communication through an intricate series of interpretations. The statement is heard, then goes through a complex sequence such as the following:

1. Objective interpretation of content (explicit message): The sender of the message thinks my dress is nice.
2. Subjective interpretation of meaning of content (implicit message): The sender doesn't really like my dress and is actually making fun of me.
3. Feeling about interpretation of meaning (angry, hurt, nervous): I feel hurt that you would make fun of my dress.

4. Feeling about the feeling (embarrassed, ashamed, inferior): I'm ashamed that I feel hurt.
5. Defense mechanism to self-protect from feeling of shame (withdraw, attack, avoid): Because I feel ashamed of feeling hurt, I'll avoid eye contact and avoid you.
6. Actual behavioral response to statement (rules or guidelines): I'll withdraw as quickly as possible, and in the future I'll go out of my way to avoid contact with you.

Obviously, the outcome of this scenario could have several different endings given a change at any step in the series. Because you make a conscious or unconscious decision at each step in the sequence based on your interpretations, it is important to be aware of your interpretations at each step. It is also important to remember that the person you're talking to is going through the same series of interpretations.

The purpose of helping relationships is to facilitate growth. The goal of communication within helping relationships is to facilitate growth through self-understanding. In Part IV of this book, therefore, we consider how to communicate to facilitate self-understanding.

13

Communicating to Facilitate Self-Understanding

You Tell on Yourself

You tell on yourself by the friends you seek, by the very manner in which you speak; by the way you employ your leisure time, by the use you make of dollar and dime. You tell what you are by the things you wear, by the spirit in which your burdens you bear; by the kind of things at which you laugh, by the records you play on the phonograph. You tell what you are by the way you walk, by the things of which you delight to talk; by the manner in which you bear defeat, by so simple a thing as how you eat. By the books you choose from the well-filled shelf—in these ways and more you tell on yourself. So there's really no particle of sense in an effort to keep up false pretense.

— ANONYMOUS

GOALS OF COMMUNICATION

Our ability to satisfy basic human needs, such as love, intimacy, and self-esteem, is in large part dependent on our ability to develop effective communication skills.

> I see communication as a huge umbrella that covers and affects all that goes on between human beings. Once a human being has arrived on this earth, communication is the largest single factor determining what kinds of relationships she or he makes with others. How we develop intimacy, how productive we are, how we connect—all depend largely on our communication skills. (Satir, 1988, p. 5)

Our ability to be effective helpers in a helping relationship is dependent on our ability to communicate effectively. Communication is the very fabric of the helping process. Communication skills are central to counselor effectiveness (Goldin & Doyle, 1991).

Communication effectiveness is determined relative to communication goals. There are three basic types of communication goals:

Type 1: To Express—the goal of this type of communication is to send clear messages (thoughts, feelings, behaviors) that can be understood by the receiver—such as lecturing, giving instructions.

Type 2: To Understand—the goal is to receive and accurately interpret messages sent (thoughts, feelings, behaviors)—such as notetaking, receiving instructions.

Type 3: To Facilitate Understanding—the goal is to help the receiver accurately interpret messages sent (thoughts, feelings, behaviors). Messages are from the receivers themselves or their interpretation of others' messages. The goal is self-understanding by the receiver of his or her own perspective—such as in helping relationships. Type 3 communication incorporates types 1 and 2.

The goal of communication in helping relationships is to facilitate understanding. To achieve this, one must also be able to express and to understand effectively.

PREREQUISITES FOR COMMUNICATING TO FACILITATE SELF-UNDERSTANDING

Because our goal is to facilitate self-understanding in our clients, our role in the communication process is different from the norm. Typically, people are most focused on having their *own* perspective clearly communicated and understood. They are secondarily focused on understanding the other person's perspective. Last, they might focus on facilitating the self-understanding of the other person. Our role in helping relationships is to communicate in such a way as to enable our clients to understand themselves better. It is not, as often misconstrued, to facilitate their understanding of *our* perspectives. It is to facilitate their increased clarity of their *own* perspectives—to facilitate their *self-understanding*.

As suggested in Part I, there are four fundamental prerequisites that will influence one's ability to facilitate the growth or self-understanding of another person. These are (1) clear purpose and goals, (2) philosophy of growth, including self-understanding and other-understanding, (3) communication skills, and (4) facilitative skills. Purpose and goals were addressed in Part I. Philosophy of growth was addressed in Part II. Self-understanding and other-understanding are differentiated below.

Self-Understanding

Self-understanding relates to one's philosophy of growth and development and applies to both client and helper. The focus of self-understanding is on the helper's understanding of self. It is personal growth. Helpers *must* be able to differentiate between their own and their clients' perspectives, issues, and growth processes. They must recognize their own personal issues well enough to prevent them from interfering with the goal of the helping process: to facilitate the growth and self-understanding of the *client*.

Other-Understanding

The focus of other-understanding is on understanding the client from the client's perspective. Other-understanding is empathy. We can be effective only when we are able to comprehend and respect the other's context and point of view in all its manifestations rather than insisting on our own. The following excerpt by Frank Koch (cited in Covey, 1990) in *Proceedings,* a magazine of the Naval Institute, serves as a reminder that insisting on our own perspective may interfere with the process of helping.

Two battleships assigned to the training squadron had been at sea on maneuvers in heavy weather for several days. I was serving on the lead battleship and was on watch on the bridge as night fell. The visibility was poor with patchy fog, so the captain remained on the bridge keeping an eye on all activities.

Shortly after dark, the lookout on the wing of the bridge reported, "Light, bearing on the starboard bow." "Is it steady or moving astern?" the captain called out.

Lookout replied, "Steady, captain," which meant we were on a dangerous collision course with that ship.

The captain then called to the signalman, "Signal that ship: We are on a collision course, advise you change course 20 degrees."

Back came the signal, "Advisable for you to change course 20 degrees."

The captain said, "Send, I'm a captain, change course 20 degrees."

"I'm a seaman second class," came the reply. "You had better change course 20 degrees."

By that time, the captain was furious. He spat out, "Send, I'm a battleship. Change course 20 degrees."

Back came the flashing light, "I'm a lighthouse."

We changed course. (p. 33)

Communications Skills

The focus of communication skills is the ability to implement effectively the basic components of communication—listening, responding, and expressing (see Chapter 14)—through the mechanisms of verbal and nonverbal behavior, statements, and questions (Chapter 15).

Facilitation Skills

Facilitation skills refer to the ability of the helper, based on self- and other-understanding, to make effective decisions regarding the *use* of communication skills to facilitate the growth (self-understanding) of the client. Facilitation skills assist a helper in making effective decisions about when to listen, respond, and express. An effective counselor can decide what type of facilitation skills to use (self-disclosure, empathy, confrontation) in responding or expressing. These skills help a counselor respond rather than react, and stay focused on client need over personal need (see Chapter 16).

PATTERNS OF COMMUNICATION

Because we all have been communicating for many years, we have developed many patterns of communication. Habits form that we continue, generally unconsciously, regardless of how effective they are, until we are confronted by a situation in which our communication skills fail us, or until we decide to do a conscious self-analysis. The self-analysis should include the following:

1. Data Collection. Collect information as objectively as possible about your patterns. One way is to set up a dyad in which you are communicating with another person while a third observer notes patterns.
2. Pattern Identification. Identify patterns based on input and assessment of your observer, communication partner, and yourself.
3. Evaluate Effectiveness. Analyze the effectiveness of your communication patterns, based on the goal of the communication process.

This self-analysis can be applied to the communication components of listening, responding, and expressing. It is important to keep in mind what your goal is and to evaluate effectiveness relative to that goal. The goal of communication in the helping process, and therefore for our purposes, is the facilitation of client self-understanding and growth.

While all types and styles of communication may well have a time and place of appropriateness, it is important to consider the impact and effect of a *pattern* developed. Consider the message given to the person with whom you're communicating, and consider whether the pattern is effectively facilitating the growth of that person through self-understanding.

CRITERIA FOR GIVING FEEDBACK

Feedback is a way of helping another person to consider changing his or her behavior. It is communication to a person or a group that gives them information about how they affect others. Feedback helps individuals keep their behavior "on target" and thus better achieve their goals. Following are some criteria for useful feedback:

1. *It is descriptive rather than evaluative.* When behavior is described rather than judged, the individual can hear the feedback nondefensively and use it as he or she sees fit. When evaluative language is avoided, the individual's need to feel self-protective is reduced.
2. *It is specific rather than general.* To be told that one is "dominating" will probably not be as useful as to be told that "just now when we were deciding the issue, you talked more than anyone else." "Dominating" could also be interpreted as evaluative.
3. *It takes into account the needs of both the receiver and giver of feedback.* Feedback can be destructive when it serves only our own needs and fails to consider the needs of the person on the receiving end.
4. *It is directed toward behavior that the receiver can do something about.* Frustration is only increased when people are reminded of shortcomings over which they have no control.
5. *It is solicited rather than imposed.* Feedback is most useful when the receiver has formulated the kinds of questions that those observing can answer.
6. *It is well timed.* In general, feedback is most useful at the earliest opportunity after the given behavior (depending on the person's readiness to hear it, support available from others, and so on).
7. *It is checked to ensure clear communication.* A helpful strategy is to ask the receiver to rephrase the feedback to verify that it corresponds to what the sender had in mind. When feedback is given in a training group, both giver and receiver have the opportunity to check with others in the group on the accuracy of the feedback. Is this one person's impression or an impression shared by others?

Feedback is a way of giving help; it is a corrective mechanism for individuals who want to learn how well their behavior matches their intentions; and it is a means for establishing one's identity.

SUMMARY

Goal clarification is important to effective communication. For our purposes, the goal of communication is facilitation of client growth through self-understanding. Prerequisites to communicating for the facilitation of growth through self-understanding include (1) clear purpose and goals, (2) philosophy of growth, including self- and other-understanding, (3) communication skills, and (4) facilitative skills.

Helpers have already developed communication skills and patterns of communicating. Self-analysis is important for the identification of patterns and their effectiveness in accomplishing the identified goal of communication. Feedback from others provides useful information regarding communication patterns, provided the feedback is given constructively.

14

Components of Communication: Listening, Responding, and Expressing

No one wants advice—only corroboration.

— JOHN STEINBECK

The process of communicating has three essential aspects: listening, responding, and expressing. These are referred to here as the basic components of communication.

Listening involves physically hearing plus mentally paying attention. All other aspects are impacted by listening. Responding means sending communication back to someone who has already communicated. Responding takes into account or focuses on the other person's perspective. Expressing involves sending a message that is focused on the sender's perspective.

This chapter addresses each of these components of communication. It introduces basic types of listening, common response styles, and typical styles of expressing. The objective of the chapter is twofold. It is first, to encourage a self-analysis of communication styles and patterns, and second, to consider whether the style and pattern promote the goal of communication: the facilitation of self-understanding.

FIVE BASIC TYPES OF LISTENING

Of the components of communication, listening is the one most people have developed least. At the same time, listening may be the most important because it affects all the others. In helping relationships, effective listening is central to understanding and to facilitating understanding.

There are five basic types of listening: nonlistening, pretend listening, selective listening, self-focused listening, and empathetic (other-focused) listening.

Nonlistening

Nonlistening means the receiver may be hearing but is not paying conscious attention to what is being said. Indications of nonlistening are inappropriate responses, interruptions, or no response.

Pretend Listening

Pretend listening looks like listening (i.e., eye contact, open body posture, nodding), but the receiver is actually nonlistening, is thinking about something else. Pretend listening can actually be quite prevalent in helping relationships when helpers feel self-conscious, are distracted or worried about what they think they "should" do, or are just focused on some aspect other than what their clients are currently saying.

Selective Listening

Selective listening refers to the receiver's screening certain types of messages or information and paying attention to others. Focusing on facts rather than feelings is selective listening. Selective listening can be a useful skill when used appropriately. The key to useful selective listening is in identifying appropriate selectiveness. The helper who selects out anything that triggers personal issues, for example, and responds defensively to them is not using selective listening effectively. On the other hand, screening extraneous noise, such as phones ringing, children playing, or distant conversations, is an example of useful selective listening. It is also with selective listening that a helper can screen through rambling narratives full of extraneous detail and select out significant aspects of thoughts, feelings, or behaviors.

Self-Focused Listening

Self-focused listening concentrates on the perspective of the listener. It involves judging, interpreting, and experiencing the information as it impacts the listener. Often self-focused listening occurs simultaneously with selective listening, as when listeners select elements that impact them, then focus on seizing an opportunity to make their own response rather than continuing to listen.

Some helpers become distracted by what they want to say and anxiously await an opportunity to interject, missing the rest of what their client is saying. They are using both self-focused and selective listening. Indications of self-focused listening include responses that are evaluative or judgmental, and rehearsing rather than listening.

Empathic (Other-Focused) Listening

Empathic listening is hearing the message and accurately understanding the *sender's* perspective. Empathy means understanding the experience of the other person, in terms of that person's thoughts, feelings, and/or behaviors.

Many people use all types of listening to varying degrees. Generally, however, they use some more than others. The types may be combined—selective, self-focused listening, or selective, empathic listening. The first four types are used more than the fifth. Ironically, the fifth type is probably the most effective in facilitating understanding.

A self-analysis of your listening style is a useful place to start when you want to improve your listening skills. Begin with an identification of your goal(s) for the communication process. Then identify the types of listening you tend to use most. Finally, observe the effects of the respective types on those with whom you are speaking. Is the listening style conducive to accomplishing your goal(s)? Specific self-analysis exercises are provided in the workbook supplement for this book.

RESPONSE STYLES

The second major component of the communication process is responding, when messages are returned *in response* to a message received. Responding is a component in which patterns and habits are readily apparent. Response styles are more or less appropriate depending on one's goal. For our purposes, the goal is facilitating client self-understanding.

Following are eight common response styles and there may be times when all styles can be used appropriately. Think in terms of how effective each might be in our goal of facilitating self-understanding.

One-Upper

The one-upper responds in a way that implies your situation isn't so significant. No matter what you say, one-uppers can top it with a better story about themselves. "You think *you* have it bad? Wait until you hear *my* situation." One-uppers reflect their perspectives rather than their clients'. One-uppers generally reflect *control* rather than an acknowledgment of *rights*.

Discounter

The discounter discredits your feelings, thoughts, behaviors, or experience. You are not taken seriously. This can take the form of put-downs ("You don't have it as bad as you think you do"), sarcasm or joking ("Life is pretty tough all over these days"), or reassurance ("I'm sure you'll feel better tomorrow"). Discounters also reflect their own perspectives. Discounters reflect *judgment* rather than *respect*.

Expert

The expert implies a hierarchy. It may be real (parent-child, employer-employee) or imposed based on the assumption that the expert "knows" more than you do. The experts say directly or indirectly that they have the answer and that your job is to do what they say: "I want you to go home, sit down, and make a list of everything that's bothering you." Experts assume authority and can be patronizing. They give commands. Experts reflect their own perspectives rather than their clients'. Experts can communicate *control, judgment,* and *rescuing.*

Advice-Giver

The advice-giver needs no hierarchy or real or implied authority to tell you what you should do. Key words for an advice-giver are "should" and "ought": "What you should do is . . . " Advice-givers demonstrate one of the more common response styles. They set themselves up to assume responsibility for their clients (rescuing) and then to be blamed when their advice doesn't work. Advice-givers can reflect *control, judgment,* and *rescuing.*

Cross-Examiner

Cross-examiners respond with a series of questions that can make people feel like they're being interviewed or cross-examined. The questions appear to be designed to get information. The underlying message with cross-examiners is that if they can get enough information, they can do something—you're not sure what. Presumably, they will then have the answer. The cross-examiner implicitly takes control of the interaction: "So did you say anything?" " Did you get upset?" Usually he or she uses closed questions. Focus is on the helper's perspective and development of perspective. Cross-examiners can demonstrate *control, judgment, rescuing,* and/or *blaming.*

"Canned" Counselor

The canned counselor reflects an insincere focus on the client's perspective. Content may focus on the client—"So how do you feel about that?"—but the underlying message suggests focus is on the helper's perspective: Once the counselor knows how the client feels, he or she will evaluate the client. "Canned" counselors can reflect *judgment* rather than *respect.*

Problem-Solver

The problem-solver responds in a way that says, "I know what your problem is or will find out." The hidden message is that you don't, or can't. The trouble with this response style is that clients never get a chance to figure things out for themselves, faced with such questions as "So have you tried talking directly to your father yet?" The problem-solver can incorporate cross-examining and advice-giving. The focus is on the helper's perspective. Problem-solvers can reflect *control*, *judgment*, and *rescuing*.

Empathizer

The empathizer focuses on trying to understand the client's perspective and on helping the person discover and understand his or her perspective. Empathic statements are used. When questions are used, they're open questions: "It sounds like you're really feeling upset about your performance and are wanting to do something to rectify it." "What options do you think you have to do that?" Empathizers focus on their clients' perspectives, and generally reflect *rights*, *respect*, and appropriate *responsibility*.

As with types of listening, many people use all the eight common response styles to varying degrees; however, they have a tendency to use one or two more often than the others. A message is communicated by one's *pattern* of responding. Do a self-analysis of your response style. Which one or two do you use most? What is the impact of the pattern on the person with whom you're communicating? What message is being communicated regarding your beliefs about rights, respect, and responsibility? Is the pattern conducive to the goal of facilitating self-understanding? See the workbook supplement for specific self-analysis exercises.

STYLES OF EXPRESSING

Expressing refers to the initiation of a message that comes from the sender's perspective. This is in contrast to a response, where the initial message is sent by someone else. It means identifying one's thoughts, feelings, or behaviors, and communicating them as clearly and accurately as possible to another person. Expressing reflects the sender's point of view. This involves at least a two-step process:

1. Self-identification of thoughts, feelings, or behaviors
2. Communication of thoughts, feelings, or behaviors to another person

Just as we have developed habitual patterns of listening and responding, we have also developed patterns of expressing. Four typical styles are described below. We tend to use all of these styles in varying degrees.

A self-analysis to learn which styles you use most and the effectiveness of each is a useful place to start when evaluating your communication effectiveness. Refer to the workbook supplement for specific exercises. Remember that effectiveness is judged based on the goal of the communication: client self-understanding.

Passive

A passive style of expressing (initiating a message) is one that does not express your perspective, thoughts, feelings, or behaviors. People who are passive ignore their own rights; repress their thoughts, feelings, and behaviors; and fail to take responsibility for themselves. A passive style reflects a lack of sensitivity to oneself. A passive style is often associated with fear.

The following categories of parameters, motivation, promotion, and consequences can be used in self-assessment of styles of expressing:

parameters: Passivity is dishonest, self-denying, disrespectful, and irresponsible to oneself.

motivation: Avoidance of perceived discomfort, fear of rejection, belief of unworthiness, undeservingness, lack of skills

promotion: Anxiety, fear, disappointment, anger, and resentment

consequence: You don't get what you need, don't take care of yourself, get angry, develop a poor self-concept, and desire to blame others.

Aggressive

An aggressive style of expressing is one that expresses one's perspective in a dominating, unnecessarily forceful or threatening manner. An aggressive style reflects a lack of sensitivity to the person to whom one is speaking. Aggressiveness reflects a lack of acknowledgment of the other's rights and a lack of respect. Aggressiveness is often associated with anger.

parameters: Aggressiveness is defensive, dominating, insensitive, and hostile.

motivation: Anger, fear of not getting what one wants, insecurity

promotion: Anger, guilt, self-righteousness

consequence: You create isolation and alienation.

Indirect

An indirect style is an expression of one's perspective, thoughts, feelings, and/or behaviors in an indirect or unclear manner. Responding sarcastically, expressing irritation nonverbally, rolling the eyes, sighing heavily, or taking out one's anger on an innocent bystander are all examples of indirect expression. Passive-aggressive, placative, and manipulative are types of indirect expression.

An indirect style of expressing reflects a lack of acknowledgment of rights, a lack of respect for both self and others, and a lack of responsibility-taking. Indirect expressing is often associated with both fear and anger, as well as sadness.

parameters:	Indirect expressing is dishonest, disrespectful, irresponsible, and difficult to interpret.
motivation:	Fear, avoidance of confrontation, belief in unworthiness
promotion:	Fear, anger, anxiety
consequences:	You may not get what you need, may blame others, and may develop a poor self-concept.

Direct

Direct expressing is clear, accurate expression of one's perspective, thoughts, feelings, and behaviors. Direct expressing is sensitive to and acknowledging of the rights, respect, and responsibility of yourself and others. Direct expressing is associated with feelings of gladness.

parameters:	Direct expressing is open, honest, straightforward, and self-enhancing.
motivation:	To accomplish goals, to actualize capabilities and potential, to be responsible
promotion:	Responsibility-taking, rights, confidence, self-respect, positive self-concept
consequences:	You have a positive self-concept and honest relationships.

One specific type of direct expression to be considered here is I-statements. They are declarative sentences that describe a thought, feeling, or other experience in a singular, first-person manner. For example, "I'm upset," "I'm excited to go," "I'm angry about not winning." I-statements can be used to describe subjective reactions, ideas, aspirations, hopes, or beliefs. A primary trait of I-statements is that they locate the feelings or concerns inside the person who is making the statement. This communicates that the feeling or reaction is "owned" by the person who is speaking.

An I-statement, such as "I feel jealous of you," places the focus of the communication on the speaker. The sentence does not simply begin with "I feel . . ." or "I think . . ." Conversely, a statement such as "I feel like you're incompetent" focuses the communication on the person hearing the statement. This is a you-focused statement rather than an I-statement.

While I-statements locate the feelings, thoughts, and behaviors with the speaker, you-statements, or you-messages, are declarative sentences that try to locate a thought, feeling, problem, or other experience inside someone else rather than inside oneself. They are used when people say things like "You make me mad when you . . ." or "You're not being fair." If a parent is tired and does not feel like playing with a child, a you-statement could be "You're being a pest." Some I-statements that could be used in this situation are "I'm tired" or "I don't feel like playing."

I-statements have a number of advantages over you-statements. According to Gordon (1970), they are less apt to provoke resistance and rebellion and they are less threatening. They help children grow and learn to assume responsibility for their own behavior, and they tend to influence children to send similar messages. They also promote honesty, intimacy, and openness in relationships.

I-statements are an effective way to bring up a problem in interpersonal relationships. They are effective because they locate the problem inside the person making the statement. Also, they communicate that the individuals who bring up the problem recognize that their view of it is a subjective belief rather than an objective fact, and this leaves room for other perceptions or definitions.

Skillful I-statements also are specific rather than general and focus attention on problems rather than personalities. Another advantage of thinking with I-statements is that they are simple concepts that can be understood by children, lay groups, and people with little education or sophistication (Burr, 1990).

SUMMARY

Listening, responding, and expressing are the basic components of communication. The five types of listening discussed are nonlistening, pretend listening, selective listening, self-focused listening, and empathic listening. Eight response styles were suggested: one-upper, discounter, expert, advice-giver, cross-examiner, "canned" counselor, problem-solver, and empathizer. The four styles of expressing that were discussed are passive, aggressive, indirect, and direct.

People tend to use all of these styles to varying degrees and in different circumstances. Most, however, have developed patterns of using one or two more than the others. Self-analysis of listening, response, and expressing styles relative to identified goals of communicating can provide useful information for helpers in developing effectiveness.

15

Mechanisms of Communication: Verbal and Nonverbal Behavior, Statements, and Questions

If anything of significance is said, somebody of consequence is bound to be offended.

━ MAHARISHI MAHESH YOGI

T he basic components of communication—listening, responding, and expressing—address *what* you are doing in the process of communication. The mechanisms of communication identify *how* you are doing it. The mechanisms are addressed in this chapter and can be identified as the following:

    I. Verbal
       A. Statements
          1. Expressive
          2. Empathic
       B. Questions
          1. Open
          2. Closed
   II. Nonverbal
       A. Silence
       B. Nonverbal behaviors
  III. Consonance

Mechanisms of communication can either encourage or discourage client talk, depending on patterns of usage. The objectives of this chapter are (1) to look at the mechanisms relative to our goal of facilitating client self-understanding, and (2) to encourage a self-analysis that includes answering the following questions:

1. What mechanisms of communication, or patterns, do I most often use?
2. What effect do these patterns have relative to my goal of facilitating client self-understanding?
3. What changes might I make to increase my effectiveness in accomplishing my goal?

VERBAL COMMUNICATION

Verbal communication is based on words. Common mechanisms of verbal communication are statements and questions, both of which can be used effectively and ineffectively. Patterns of usage are more significant than single-time usage.

Statements

Statements are messages to another person that do not directly request a response. Statements can be expressive, reflecting your perspective, or empathic, reflecting your client's perspective.

Expressive Statements

Expressive statements present the speaker's perspective. They are used when you want the other person to understand your point of view. Examples of expressive statements include the following:

1. *Information-Giving*—Providing factual content information to your clients that they do not know: "The domestic abuse survivors' shelter is located at 210 Elm Street. It's open 24 hours a day. You can call this phone number to obtain transportation there."
2. *Self-Disclosure*—Offering information about yourself that your client wouldn't otherwise know. Appropriate self-disclosure always focuses on the need of the client, even though it reflects the

helper's perspective: "I had a similar experience, and for me, I struggled with feeling a great deal of anger toward our legal and justice system. I'm wondering if you're experiencing any anger at our imperfect and sometimes seemingly unfair system."

3. *Immediacy*—Providing your perspective on some element of the relationship with your client: "It seems to me that in our relationship, we've developed a pattern in which you bring up questions you have and then I answer them. I wonder if that's your perspective."

4. *Summarizing*—Providing a summary, from your perspective, of the points and/or interaction so far: "What I've heard to this point is that, seeing no other choice, you left your husband, moved in with your sister, have been looking for a new job, and only now are starting to have second thoughts about whether you left too quickly."

Expressive statements can be either appropriate or inappropriate to your goal. For example, the following might be an appropriate expressive statement using self-disclosure: "I think I can relate to what you're feeling. I lost a child myself in a car accident. For me, I struggled with feelings of guilt, and I'm wondering if that's part of your struggle." The "expert," "advice-giver," and "discounter" are examples of potentially inappropriate usage. For example, "I know what you mean; I experienced the same thing. What you need to do is . . ." (advice-giver); "Stop focusing on it" (expert); "I know it's really hard, but it'll get better" (discounter).

Empathic Statements

Empathic statements reflect your understanding of the other person's perspective. They are used when you want the other person to know that you understand his or her point of view. For example, "It sounds like this is a really challenging situation for you right now, and you are very frustrated because you don't know what to do."

Examples of empathic statements include the following:

1. *Paraphrasing.* Paraphrasing is a rewording of what the client has said, reflecting the client's perspective.
   Client: "I don't think he's going to change in a million years."
   Paraphrased response: "So what you're saying is you don't think your husband will ever change."

2. *Explicit empathy*. Communicating an *understanding* of what your clients have *explicitly* stated, regarding their thoughts, feelings, or behaviors.

   Client: "It just pains me that Cloe's not going to go to college." *Explicit empathy response:* "So you're feeling really disappointed as you're realizing your daughter is not planning to go to college."

3. *Implicit empathy*. Communicating an understanding of what your clients have stated only implicitly; responding to the underlying meaning or message.

   Client: "I don't know, I always did the best I could but, you know, I haven't been a perfect father . . ." *Implicit empathy response:* "So it almost sounds as though you're starting to wonder if you might have had some influence on your son's decision."

4. *Confrontation*. Communicating an empathic understanding of the discrepancies in the client's perspective.

   Client: "I want my daughter to have fun and have a good time, but when it comes to school, she had better hit the books." *Confrontational response:* "So, on the one hand, while you want your daughter to have fun, on the other hand, you don't want to interfere with studying."

Questions

Questions are messages sent to another person that directly request a response. Questions, like statements, can be effective or ineffective in encouraging productive client talk. A major potential pitfall with questions is the unspoken message that *helper* understanding is the goal rather than *client* self-understanding. Implicitly, the assumption is that the helper is responsible for resolving the client's issue. Of the two types of questions, open and closed, open questions lend themselves to greater focus on client perspective, rights, respect, and self-responsibility.

Open Questions

Open questions are those that can't be answered in a few short words. They provide wide choice, encouraging others to talk and provide you with maximum information. Typically, open questions begin with *how, what,* or *could:* "Could you expand on that?" "What would be an example of that?" "How was your last experience in counseling?"

Open questions cannot be answered "yes" or "no," and do not supply a set of answer options. They allow clients freedom to explore and take the interview where they wish.

Open questions are particularly useful in the exploratory phase when you may not know the kinds of specific narrow questions appropriate for the respondent's experiences (Gorden, 1992).

Closed Questions

Closed questions can be answered in few words, such as "yes" or "no." They provide restricted choice. They have the advantage of focusing the interview and bringing out specifics, but they place the primary responsibility for talk on the helper. Closed questions often begin with *is, are,* or *do:* "Is your father living?" "Did you benefit from your experience in counseling?"

Before using closed questions, the helper must understand the topic well enough to know most of the possible answers; the answer categories must be relevant to the central purpose of the interview and be clearly meaningful (Gorden, 1992).

Note that a question, open or closed, on a topic of deep interest to the client will often result in extensive talk-time *if* it is interesting and important enough. If an interaction is flowing well, the distinction between open and closed questions is less important.

Effective Uses

Questions have some particularly effective uses, as shown in the examples below:

1. Beginning the session/interaction: "What would you like to talk about today?"
2. Gathering information for assessment of the situation: "What can I do for you?" "Why are you seeking counseling?" "How long has this been happening?"
3. Clarifying general statements and unclear terms: "What do you mean by your 'life's falling apart'?" "When you say you made a fool of yourself, what does that mean?"

Potential Problems

There can also be pitfalls in the use of questions:

1. *Questions and rights versus control:* Questions can be controlling. Helpers who take the position of asking a series of questions put themselves in a position of controlling the session, focus, and direction: "So what'd you do next?" "And then what'd you do?" "Weren't you scared?"
2. *Questions and respect versus judgmentalness:* Questions can reflect judgmentalness. They can be framed as disguised statements of judgmentalness: "Don't you think you are a little hard on him?" "Don't you think it would help if you apologized?"
3. *Questions and appropriate responsibility-taking versus rescuing:* A pattern of questions (cross-examining) can put the client on the defensive. It can suggest that the helper is in charge and is there "to diagnose, prescribe, and fix" the client: "Were any other people present?" "Did you talk with them?" "Did you end the interaction?" "Why did you talk to them?" "What did you do with the information?"

NONVERBAL COMMUNICATION

Verbal communication consists of words. Nonverbal communication consists of everything else that results in communication between people. Nonverbal communication includes silence and nonverbal behaviors.

Silence

Silence, for our purposes, refers to being verbally quiet and nonverbally neutral. Silence is useful when productive processing is occurring for the client.

Nonverbal Behaviors

Nonverbal behavior refers to all behavior, other than verbal, that results in the transmission of messages from one person to another. Typically, nonverbal behaviors include facial expression, body language, voice tone, gestures, and eye contact. Nonverbal behavior is generally unconscious and is therefore often more significant and accurate than verbal behavior. Gorden (1992, p. 66) cites studies suggesting that as much as 65% of the meaning of a spoken message is determined by nonverbal communication.

> Most non-verbal behavior is automatic and unconscious and, therefore, more difficult for either the interviewer or the respondent to consciously control than verbal behavior. . . . Clients in counseling sessions are more aware of their *words* than of their non-verbal behavior.

When verbal and nonverbal communication are inconsistent, we generally respond to the nonverbal. For example, when a friend says, "Thanks a lot" with an intonation of sarcasm, the nonverbal communication reflects a *lack* of appreciation, countering the verbal content of appreciation. When such inconsistency arises, we're more apt to trust the nonverbal. How something is said can be as important as what is said (Phares, 1992).

Because nonverbal behavior is generally deferred to as a more accurate reflection of meaning and because the sender of the message is generally unaware of his or her action, the helper must pay particular attention to becoming conscious of patterns of nonverbal behavior. A self-analysis, using feedback from others or videotaping, can be helpful in identifying nonverbal behavior patterns, their effect on others, and their impact, if any, on the goal of facilitating understanding. It is imperative to remember that nonverbal behavior reflects diverse individual, family, and cultural backgrounds. Below are examples of the more common nonverbal behaviors.

Body Position

Body position includes proximity to the other person and posture—open stance, head erect, slight forward lean.

Gestures and Body Movements

Both voluntary movements, such as hand gestures, arm gestures, and head nodding, and involuntary movements, such as twitches and eye blinking, are included here.

One's body language gives especially useful cues to that person's feelings—including moods, mental states, and physical comfort or discomfort—that provide information about the client's behavior that can be investigated more specifically through talk. (adapted from Hutchins & Cole, 1992)

Facial Expressions

Voluntary expressions, such as smiling, frowning, eyebrow raising, lip pursing, and involuntary expressions, such as blushing or paling, can carry powerful messages.

Facial expression is the major means of communication, next to human speech. (adapted from Hackney & Cormier, 1988, p. 32)

Eye Movement

Eyes are very expressive. Nonverbal behavior includes both degree of eye contact and expression—glaring, squinting, tearing, pupil dilation.

When a respondent makes direct eye-to-eye contact in response to a question or statement, he or she is usually indicating a willingness to communicate, attentiveness, and a desire to understand. (adapted from Gorden, 1992)

Voice

Voice refers to tone, inflection, volume, quality, breath, breathing, and timing of statements.

Tone of voice and inflection tell the client whether the counselor is accepting, even if he or she uses [conceptually meaningless vocalizations such as "mm" instead of words]. (Brammer, 1988, p. 64)

The use of well-modulated, unexcited vocal tone and pitch will reassure clients of your own comfort with their problems. (Hackney & Cormier, 1988, p. 35)

Nonverbal behaviors can be used specifically as minimal encouragers for the client. Encouragers are verbal and nonverbal responses that prompt the client to continue talking. They are generally simple head nods, facial expressions reflecting interest or concern, or perhaps minimal sounds, such as "Hmm," which encourage the client to continue talking (Ivey, 1994).

CONSONANCE

Consonance refers to congruence of verbal and nonverbal messages. It reflects a consistency among thoughts, feelings, and behaviors, as demonstrated by verbal and nonverbal behavior. Consonance facilitates accurate interpretation of messages. For example, a statement such as "I'm hurt," when accompanied by tears, emphasizes the verbal content and is believable. The same statement, said while smiling, communicates mixed messages: "I'm hurt" and "I'm not hurt." Mixed, or double, messages lack consonance and are more challenging to interpret.

SUMMARY

Basic categories of mechanisms of communication include verbal, non-verbal, and consonance. Verbal mechanisms include statements and questions. Statements can be expressive—information-giving, self-disclosing, having immediacy, summarizing—or empathic—paraphrasing, showing explicit empathy, showing implicit empathy, and confrontational. Questions are open or closed. Nonverbal mechanisms include silence and nonverbal behaviors (body position, gestures, facial expressions, eye movement, and voice).

Mechanisms of communication can be used either effectively or ineffectively. They can either encourage or discourage client disclosure.

16

A Five-Step Communication Model

"You're wrong," means "I don't understand you"—I'm not seeing what you're seeing. But there is nothing wrong with you, you are simply not me and that's not wrong.

— HUGH PRATHER

Y ou can use words either to express your perspective or to help others understand their perspective better. As our goal in helping relationships is to facilitate growth and client self-understanding, the second of these is more effective most of the time (Thompson & Stroud, 1984).

Certain roles, such as a helper role, require that we set aside our personal perspective, feelings, judgment, preference, or opinion and respond *purposefully* to our client. The point is to remain personally detached and to communicate for the express purpose of facilitating growth and self-understanding of the client.

Staying personally detached means being *able* to set aside your feelings and opinions, and to stay focused on the goal. This can be quite challenging, particularly if your client triggers reactions in you by such behaviors as expressing anger with you, criticizing you, challenging you, or making decisions contrary to your values.

The model for communicating presented in this framework provides a five-step framework from which to communicate. The model incorporates the three basic communication components of listening, responding, and expressing. It reflects the three R's of rights, respect, and responsibility. It also provides an umbrella for the effective helping prerequisites of self-understanding, other-understanding, communication skills, and facilitative skills. It allows the helper to assume a position for communicating effectively in even the most challenging circumstances. The model translates principles from the martial arts and applies them to communication.

MARTIAL ARTS: A FOUNDATION FOR COMMUNICATION

The martial arts evolved for the purpose of helping a person survive in a tough world. They were developed to respond to the danger of physical assault. Less dangerous but nonetheless painful are attacks on our sensibilities, beliefs, and feelings for others, which are also vulnerable to attack. How do we survive in this kind of world? We can try fighting fire with fire, but that usually just creates a bigger fire. And even if we win some battles, there's sure to be someone or something that will eventually frustrate our efforts and beat us. The philosophy of martial arts is to build yourself up from the inside out, to develop strength, suppleness, sensitivity, coordination, and confidence. Martial arts is based on self-respect, mirrored in respect for others. To neutralize an attack without causing injury requires skill and ethical intention. It enables you actually to nurture an aggressor without harm to yourself or others (Thompson & Stroud, 1984). As helpers, we often need to apply these same principles through our communication. Communicating an attitude of rights, respect, and appropriate responsibility-taking toward both ourselves and our clients reflects the martial arts philosophy of protecting your attacker and yourself from harm.

The martial arts are based on a principle of harmony and unity with nature. *Judo* means the soft or gentle *(ju)* way *(do)*. *Aikido* means a way *(do)* of uniting the mind with nature *(aiki)*. A basic philosophy in martial arts is to bring energies within the self, or between the self and others, into harmony by clarifying and honoring these energies.

Martial arts, to be effective, must incorporate the following five steps:

1. Observe
2. Balance

3. Align
4. Direct
5. Follow Through

Observe

To observe means to pay attention to what's happening. If another person is coming toward you with the potential intent to harm, you need to watch and observe. Often people want to believe that if they don't look and don't see, something doesn't exist. If you don't look and you don't observe, you simply put yourself at a disadvantage in dealing with the situation.

In communication, if you don't pay attention, look, and observe, you not only are at a disadvantage, you may not be able to proceed with the communication. Paying attention in communication means listening. If you don't listen, you cannot respond effectively.

Balance

As your opponent comes toward you and makes physical contact with you, you must balance yourself in order not to be pushed over. Being balanced means your body weight is over your feet, your knees are slightly bent, and you are positioned to absorb movement and impact on your body. The point is to position yourself to act appropriately without sacrificing your own balance, your own internal equilibrium.

In communication, you need to be emotionally balanced, or centered, in order to respond effectively. If you're off balance emotionally, you're more likely to respond defensively than effectively. Again, you need to maintain an internal equilibrium.

Align

To align, physically, means to put your body in motion in the same direction in which the other person is moving. As an opponent makes contact against your body, you turn and move with his or her momentum. As a martial art, aikido's purpose is to render an attack harmless, not by opposing it with force, but by accepting it, aligning oneself with the direction and energy of the attacker, and letting that same energy defeat itself. The aggressor, finding nothing there, pulls himself or herself off balance (Leviton, 1992).

In communication, when you align initially with what the speaker says, you demonstrate that the person is being heard and understood. In communication, aligning is empathizing.

Direct

Once you are physically aligned and going in the same direction as your opponent, you are in a position to direct that person's movement. The directing may involve a variety of moves or skills on your part. The physical aligning with the opponent's body momentum, however, is a prerequisite to the ease and effectiveness of the directing moves.

Directing in martial arts is equivalent to focusing in communicating. Focusing provides for greater and greater clarification.

Follow Through

Just as a hitter follows through with a baseball bat, the follow-through in martial arts provides power, direction, and accuracy. Follow-through in communication is what provides for direction and a sense of closure.

_____

FIVE-STEP COMMUNICATION MODEL

The five steps used in the martial arts can be translated into five equivalent communication skills that apply directly to the communication process, as shown below.

FIVE-STEP MARTIAL ARTS MODEL AND
COMMUNICATION SKILLS EQUIVALENTS

MARTIAL ARTS:	*COMMUNICATION SKILLS:*
1. Observe	1. Listen
2. Balance	2. Center
3. Align	3. Empathize
4. Direct	4. Focus
5. Follow Through	5. Provide Directional Support

These steps, when applied to communication, are presented as a five-step model for effective communication in facilitating self-understanding. The steps are sequential. You can't center until after you've listened. You may be able to center, but not *in relation to* what you're going to hear. You can't empathize effectively unless you're centered; you don't focus until after you've empathized; and you provide directional support *after* you have focused. You do, however, go back and forth in the model. With every new statement and/or concept, you go through the sequence again. The steps apply to individual statements, individual sessions, and overall direction. The five steps are discussed below.

Listen

Applied to the communication process, to observe means to listen. Our goal in communicating is to facilitate understanding and self-understanding. A prerequisite to understanding is listening.

The following example from Covey's (1990) *The Seven Habits of Highly Effective People* points out the importance of listening to understanding:

> A father once told me, "I can't understand my kid. He just won't listen to me at all." "Let me restate what you just said," I replied. "You don't understand your son because he won't listen to *you*?" "That's right," he replied. "Let me try again," I said. "*You* don't understand your son because *he* won't listen to you?" "That's what I said," he impatiently replied. "I thought that to understand another person, *you* needed to listen to *him*?" I suggested. "Oh!" he said. There was a long pause. "Oh!" he said again, as the light began to dawn. "Oh, yeah! But I do understand him. I know what he's going through. I went through the same thing myself. I guess what I don't understand is why *he won't listen to me*." (p. 239)

Listening is a combination of attending, hearing, interpreting, and ultimately understanding. As the intent of listening is to understand, effective listeners use all sources of communication—verbal and nonverbal—in their efforts to accurately understand.

When the process of listening is analyzed, it includes the following steps:

1. Attending—paying attention, concentrating on the communication.
2. Hearing—the physical process of using your auditory sense. You can hear without paying attention, but you won't understand. You can also pay attention but be unable to hear.
3. Explicit comprehension—hearing and comprehending at face value what is said; hearing and decoding by assigning meaning to the verbal message.
4. Implicit comprehension—decoding the nonverbal message, the statement between the lines.

> It is as important to attend to what is omitted by the client as it is to hear what is actually being said. (adapted from Horney, 1945)

5. Understanding—reading the combined verbal and nonverbal message.

> Effective listening means hearing the clients' internal meanings and understanding their personal worldview. (adapted from Beck & Weishaar, 1989)

> Listening in its deepest sense means listening to clients themselves as influenced by the contexts in which they live, move, and have their being. (Egan, 1994, p. 98)

6. Remembering—recognition as a result of concentrated focus on hearing, comprehension, and understanding, using auditory, verbal, and nonverbal sources.

Listening is a prerequisite to the use of all other communication components and helping skills for effective communicating. Specific skills relative to listening are presented in Part V.

Center

Centering is the emotional equivalent in communication to balancing physically in martial arts. Centering means listening to yourself. It

NO MATTER WHAT IS GOING ON FOR THE CLIENT, IT IS ESSENTIAL FOR THE COUNSELOR TO BE CENTERED.

means self-awareness. It means becoming conscious of all the emotional/mental dynamics occurring within you and addressing them in such a way as to allow you to keep your balance emotionally. If you are not centered during an interaction, the other person can trigger issues in you to which you react rather than remaining focused on your goal.

> The difficulty is that when we're in the fire of conflict, most of us go into an automatic patterned reaction—we fight, flee, or freeze rather than center and blend. To break these habits, we need to rigorously learn and practice new skills, particularly how to be calm, centered, and connected while under pressure.
>
> The aikidoist attempts to execute techniques with a precise balance of vigor and calm, assertion and acceptance, self-control and spontaneity—a balance which is ever shifting moment to moment. (adapted from Leviton, 1992)

Centering means listening to yourself. It means being self-aware, able to differentiate personal issues from client issues, and able to take responsibility for your own issues. The ability to center is a necessary step to being an effective helper.

To be an effective helper, you need to listen not only to the client but to yourself. You don't want to become self-preoccupied and stop listening to the client, but listening to yourself can help you identify what is standing in the way of your being with and listening to the client. It is a positive form of self-consciousness. (adapted from Egan, 1990)

Counselors remain optimally in a state of high awareness, a state in which they are operating from a holistic rather than a fragmented self (adapted from Raskin & Rogers, 1989). This centering is ongoing and creates the possibility of genuine empathy.

[The counselor] is in touch with his feelings and is able to express them and to accept responsibility for them. (Davison & Neale, 1986, p. 486)

Although the term *centering* is seldom used by proponents of the major counseling orientations, virtually all of them refer to the importance of the elements of centering to counseling effectiveness. Being self-aware, differentiating between personal and client issues, and addressing personal issues to prevent impact on the counselor-client relationship are alluded to in discussions of psychodynamic, humanistic, cognitive, and behavioral approaches.

The therapist invites transference as part of the process and must also be fully aware of countertransference. Therefore, it is crucial for therapists to be in touch with their own issues and feelings as they sort out the dynamics of their clients. (adapted from Hall & Lindzey, 1985)

The therapist is fully and accurately aware of what he is experiencing at this moment in this relationship. It allows the counselor to relate with genuineness. (Rogers, 1961, p. 282)

Cognitive therapy operates under the notion that therapists must be clear within themselves as to what are indeed irrational beliefs. They are assumed to be able to make such distinctions; thus, they must be aware of this within themselves. (adapted from Carmin & Dowd, 1988)

The behavioral therapist is a significant role model for the client; thus the therapist must be aware of the role he or she plays in the client's learning process. (adapted from Corey, 1991)

Empathize

In order to help or to interact effectively with another, you first need to understand that person. You can understand anybody as long as you can see his or her perspective. Consider the following example from Stephen Covey's (1990) *The Seven Habits of Highly Effective People:*

Suppose you've been having trouble with your eyes and you decide to go to an [opthalmologist] for help. After briefly listening to your complaint, he takes off his glasses and hands them to you. "Put these on," he says. "I've worn this pair of glasses for ten years now and they've really helped me. I have an extra pair at home; you can wear these." So you put them on, but it only makes the problem worse. "This is terrible!" you exclaim. "I can't see a thing!" "Well, what's wrong?" he asks. "They work great for me. Try harder." "I am trying," you insist. "Everything is a blur." "Well, what's the matter with you? Think positively." "Okay, I positively can't see a thing." "Boy, are you ungrateful!" he chides. "And after all I've done to help you!" What are the chances you'd go back to that [opthalmologist] the next time you needed help? Not very good, I would imagine. (p. 236)

In communication, aligning with the momentum of the other person is called *empathizing*. Empathizing means looking at the situation from the other person's perspective. It does not mean expecting that person to look at it through your eyes. Empathizing literally means aligning with the other person's experience—feelings, thoughts, or behaviors.

In *The Art of Counseling,* Rollo May (1989) defined empathy by analyzing its etymological roots. According to May, the word *empathy* has
two components: *em,* which means "in," and *pathos,* which means "a
strong and deep feeling." Empathy, therefore, can be defined as "feeling
into" the reality of another (p. 61). Empathy is a basic necessity for all
forms of genuine interpersonal encounter, including counseling (May,
1989).

Whereas attending and listening are the skills that enable helpers to
get in touch with the world of clients, empathy enables them to understand their world. Empathy has two definable elements: the recognition of another's feelings and the response of understanding it. Empathy uses the communication component of responding. Empathy is
responding to others with understanding of their perspective.

Empathy has been defined as the ability to tune into the client's
feeling and to be able to see the client's world as it truly seems to the
client. An empathic response or expression communicates an understanding of the client's frame of reference and accurately identifies the
client's feelings (George & Cristiani, 1990).

[Empathy is the ability of the helper to respond] sensitively and
accurately to the client's feelings and experiences as if they were
his own. (George & Cristiani, 1990, p. 130)

[Empathy involves] translating your understanding [of the
client's thoughts, behaviors, and/or feelings into a response
through which you share that understanding with the client.
This communication is basic empathy.] If your perspectives are
correct, then it is accurate empathy. (Egan, 1990, p. 129)

Seek first to understand, then to be understood. (Corey, 1991,
p. 235)

Empathy means aligning with the client's perspective, regardless of
what it is. Remember that empathizing is *not* the same as agreeing, condoning, or encouraging. The client may be expressing anger toward
you or practicing other forms of resistance. Align with the resistance.

Resistance is anything that gets in the way of helping. Clients can
bring resistance to the helping relationship and also have it activated

during the helping process. It can be present at any stage. At best, most clients are ambivalent when they come for help. At the same time that they want change, they may have anxieties both about changing and about the helping process—for instance, the need to disclose their weaknesses. "Helpers may wrongly attribute the sources of clients' resistances, and blame them for lack of cooperation and progress" (Nelson-Jones, 1993, p. 160).

Avoid giving in to or fighting back against resistance. Instead, empathize. See the event through your client's eyes, even if the person is prone to violence and you don't agree with his or her perspective of the event. All of us believe we're acting reasonably; we want to be understood and have our perspectives acknowledged. For some clients, in their minds and in the context of the events that they know, their perspective is the *only* perspective. They must feel that they are being understood *before* they are ready to consider any other perspective. They need validation and affirmation. Let them see themselves the way they want to. *Then* you can redirect their focus and invite them to look at the full consequences of their perspective and all their options. If they don't feel threatened by you, they will ultimately *want* to broaden their perspective and be ready to redirect their focus.

Focus

Focusing, in communication, means to help direct the attention of the interaction in such a way as to have the meaning (self-understanding) become more and more clear. Just as the focus on a camera lens clarifies the image, focusing in communication clarifies and provides for understanding.

Focusing is a skill that involves directing the attention in the interaction to certain aspects, points, or directions. Focusing is at the core of the helping relationship. It is through focusing that underlying issues can be identified, examined, and reassessed, and new perspectives developed.

Focusing can be accomplished with both (1) empathic statements and (2) expressive statements. Empathic statements—statements verbalizing your understanding of the client's perspective—can focus on client feelings, thoughts, or behavior. Empathic statements can use such skills as empathy and confrontation. Expressive statements verbalize *your* perspective in providing focus; they can use such skills as immediacy or self-disclosure (see Part V).

Provide Directional Support

Providing directional support in communication means providing enough follow-through to lead to resolution or closure in the interaction or helping relationship. Once issues are identified and examined, and the preferred outcome or action is acknowledged, you continue to provide support as your client "walks through" the preferred scenario. Directional support might take the form of talking through imagined outcomes, or meeting over a period of time while the client actually works through a behavioral change.

SUMMARY

A five-step communication model that can be used to facilitate effective communicating includes the following steps:

1. Listen
2. Center
3. Empathize
4. Focus
5. Provide Directional Support

Specific skills to facilitate the accomplishment of each step are presented in Part V.

Things which matter most, must never be at
the mercy of things which matter least.

—— GOETHE

Facilitative skills are communication skills used effectively for the purpose of promoting growth. A structure of stages and skills is a systematic, operational organization of facilitative skills.

The real voyage of discovery consists not in seeking new landscapes, but in having new eyes.

— MARCEL PROUST

R egardless of theoretical approach, one needs a process structure from which to operate in a helping relationship. "Process" refers to the actual interaction with clients, and "structure" means one's mental organization and conceptualization of the process.

The stage structure presented in Part V includes these elements:

1. Three stages reflecting a beginning, middle, and end
2. Stages that are cumulative (Stage 2 incorporates Stage 1, Stage 3 incorporates Stages 1 and 2) and hierarchical (Stage 1 is a prerequisite for Stage 2)
3. A foundation of process and outcome goals
4. Goals for client growth, helper skills, and relationship goals (see Preface Table)

Helper skills are presented and integrated into the five-step communication model that was introduced in Part IV.

The stage structure has three stages, and each stage is discussed in terms of (1) client growth (process and goals), (2) helper skills (steps of communication model and specific communication skills), and (3) relationship goals. The stages are not time dependent but process dependent. In other words, certain processes rather than a set amount of time are needed for each stage. Stage 1, therefore, may last weeks with one client and literally minutes with another. The point is to achieve the outcome goal of trust and rapport in the relationship, regardless of how long it takes. Movement from Stage 2 or 3 back to an earlier stage is also possible, particularly as clients bring in new issues and material.

The structure is eclectic and can be used with a variety of counseling approaches. It is particularly compatible with humanistic principles in Stage 1, behavioral principles in Stage 3, and humanistic, cognitive, behavioral, and psychodynamic principles in Stage 2.

Following is a summary of the assumptions underlying this structure:

1. The purpose of the helping relationship is to facilitate the process goal of growth and outcome goal of psychological health, as evidenced by a positive self-concept.
2. The process goals for the client, helper, and helping relationship are growth, facilitation of growth, and an environment conducive to growth, respectively.
3. Growth is defined as movement toward psychological health as evidenced by a positive self-concept.

4. A positive self-concept is defined as
   a. an affective experience of
       self-acceptance
       self-esteem
       self-actualization
   b. a behavioral experience of
       congruence
       competence
       internalized control
   c. underlying cognitive beliefs of
       rights
       self-respect
       self-responsibility
5. Individuals have the right to be themselves—short of harm to others—have unique abilities and differences, and have the responsibility for their own actions and life choices.
6. Inherent in individuals' right, respect for, and responsibility to be themselves is an acceptance and respect for their unique cultural, ethnic, racial, spiritual, gender-role, familial, and individual perspectives.
7. The structure is used in conjunction with the ethical guidelines of the American Counseling Association (ACA; 1988) and the American Psychological Association (APA; 1981).
8. The structure itself is eclectic and can be used with a variety of theoretical approaches.

The three chapters in Part V will address consecutively each of the three stages in the stage structure. Each stage is discussed by exploring the (1) client process, (2) helper skills, and (3) helping relationship goals (see Part V Table).

The *client process* has a process goal of growth. Outcome goals include (1) cognitive beliefs of the right to be oneself, self-respect, and self-responsibility; (2) behavioral goals of congruence, competence, and internal control; and (3) affective experience goals of self-acceptance, self-esteem, and self-actualization.

The *helper skills* have a process goal of growth facilitation. Outcome goals include (1) attitudinal beliefs of the right to be oneself, respect, and responsibility; (2) behavioral goals of genuineness, positive regard, and focusing (empathic and expressive); and (3) affective experience goals of a role model, catalyst, and facilitator.

PART V TABLE A Stage Structure for the Helping Process

Stage Focus	Client Process	Helper Skills	Helping Relationship Goals
Pre-Stage *Assumptions:* Underlying Beliefs and Attitudinal Goals	Underlying Beliefs: Rights Respect Responsibility	Attitudinal Goals: Role Modeling 3 R's Rights Respect Responsiblity	Resulting Rapport: Rights Respect Responsibility
Stage 1: Presenting Problem	Telling the Story Initial Awareness • Thoughts • Feelings • Behavior Initial Problem Identification	*Model Step* 1. Listen 2. Center 3. Empathize — *Skills* Attending Genuineness Positive Regard Boundary Distinction Explicit Empathy Concreteness	Rapport and Trust 3 R's Stage 1 Client Process Stage 1 Helper Skills
Stage 2: Underlying Issues	Differentiating the Problem • External Disappointments • Internal Issues Examining the Issues Developing New Perspectives and Insights	4. Focus — Empathic: Implicit Empathy Confrontation Expressive: Self-Disclosure Immediacy	Processing and Understanding 3 R's Stage 1 and 2 Client Process Stage 1 and 2 Helper Skills
Stage 3: Direction and Change	Direction Implementation Change	5. Provide Directional Support — Directionality Implementational Support	Directionality and Change 3 R's Stage 1, 2, and 3 Client Process Stage 1, 2, and 3 Helper Skills

The *helping relationship* has a process goal of establishing a growth-conducive environment. Outcome goals for the helping relationship include (1) acknowledged beliefs of rights, mutual respect, and appropriate responsibility-taking; (2) behavioral goals of rapport, effective processing, and directionality; and (3) affective experience goals of trust, understanding, and change.

17

Stage 1:
The Presenting Problem

*I would not dream of belonging to a club
that is willing to have me for a member.*

— GROUCHO MARX

Stage 1, or the beginning stage, includes the period of time from initial client contact through initial problem identification as well as trust and rapport building in the relationship (See Part V Table). This may involve weeks or minutes. It is the time the presenting problem—the reason the client is there—is addressed. The presenting problem might initially be identified by the client or, as in the case of a resistant or referred client, by someone else, such as a parole officer, the client's partner, or the client's parent.

The client process, in Stage 1, focuses on the presenting problem. It includes (1) telling the story, (2) initial awarenesses (thoughts, feelings, behaviors), and (3) initial problem identification. The initial problem identification is generally an external focus: an inattentive parent, irresponsible child, insensitive boss, loss of a job.

Helper skills in Stage 1 begin with the attitudinal goals of rights, respect, and appropriate responsibility-taking. They then incorporate steps 1–3 of the communication model: listening, centering, and empathizing. The specific skill of attending is used to promote effective listening. Centering is facilitated through the use of genuineness, positive regard, and boundary distinction. Empathy is developed through use of the specific skills of explicit empathy and concreteness.

Relationship goals are based on the three R's: rights, respect, and appropriate responsibility-taking, as applied to the helping relationship. Goals in Stage 1 include rapport and trust.

CLIENT PROCESS

The client process in Stage 1 includes telling the story, initial awareness, and initial problem identification. These three processes should illuminate the presenting problem.

Telling the Story

Telling the story refers to clients' presenting their perspective on what's wrong and why they are there. For each client, it is his or her story. Clients need the opportunity to tell and to be heard. Their storytelling may include both narrative content and emotional catharsis. The narrative content is the factual information about what they are experiencing. The emotional catharsis is the unburdening of emotions clients feel are related to their circumstances. Accumulated emotions, such as fear, hurt, sadness, and relief, often emerge through their storytelling as they talk to someone they feel will understand.

> Clients need to reveal and discuss their problem situations and their missed opportunities in order for helpers to be of service. (adapted from Egan, 1990, p. 29)

> In gathering information about the nature of a client's difficulties, it is important to ask for specific information from the client. (adapted from Brammer, 1988, p. 70)

Initial Awareness

Clients' initial awareness of their feelings, thoughts, and behaviors is important for clarification of the presenting problem.

Feelings

Following are common ways clients tend to deal with feelings:

1. Withholding feelings. Keeping feelings inside and giving no verbal or nonverbal cues.
2. Expressing feelings. Giving an immediate verbal or nonverbal cue that reveals their feelings, including open expression of positive and negative feelings.
3. Describing feelings. Putting the immediate emotional state into words. This is probably the best strategy for dealing with feelings, though it is a separate issue from controlling behavior. To describe feelings, clients need to
   a. experience and identify their feelings
   b. be specific
   c. indicate what has triggered the feelings
   d. make sure that the feelings are theirs (based on Verderber & Verderber, 1989, pp. 161–167).

Thoughts

Undoubtedly the greatest power we and our clients have is the ability to change our minds, our perspectives, our thoughts. Probably the most significant change we can make is inside our heads. Events are out of our control and always in flux. The only thing we can actually control is how we perceive the events and the experience we have as a result of how we judge them.

What we focus on in our heads becomes, in fact, our reality. Our underlying cognitive beliefs, therefore, are perhaps the single most important influence on how we experience ourselves and our lives. We take responsibility for our lives by taking responsibility for our thoughts, both the content and process. Fortunately, thoughts are within our control.

Behavior

The model shown in Figure 17.1 is useful in helping clients understand impacts on their behavior. It demonstrates the relationship among feelings, thoughts (beliefs), and behavior. The following anecdote illustrates how the model can be applied.

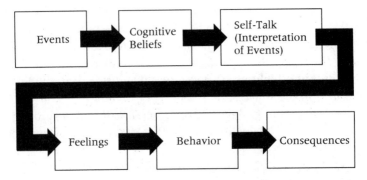

FIGURE 17.1 Relationship of events, beliefs, behavior, and consequences

The client's boyfriend goes backpacking and doesn't invite her (event). The client holds a belief that if her boyfriend really loved her, he'd want her to be with him all the time (cognitive belief). Resulting self-talk includes "I must not be a good enough hiker." "My boyfriend must not love me." The self-talk results in the client's feeling hurt (feelings). When she feels hurt she withdraws and sulks (behavior). This results in the boyfriend's being less interested in spending time with her (consequences). Of course, the event, beliefs, self-talk, feelings, behaviors, and consequences all include an endless number of possible variations, which in turn change subsequent sequences.

Initial Problem Identification

Initially clients tend to identify the source of their problem as external. The problem is usually somebody, something, some event, some circumstance:

My husband won't listen to me.
My kids are irresponsible.
My mother is always on my back.
My house is too small.
My car is too old.
My father treats me like a child.

Clients often want to spend more time describing the problematic behavior of others than their own experience of these problems, and how someone else's behavior is a problem for them.

When the helping relationship begins with these externalizing complaints about the behavior of others, the helper needs to respond to clients' immediate experience, but this is only the first step (Teyber, 1992). Ultimately, of course, the initial external problem needs to be differentiated from the internal issue, or why this particular external situation is a problem for the client. This differentiation will then be explored in Stage 2.

HELPER SKILLS

In looking at specific skills, it is imperative to keep in mind their purpose. The helper skills used in Stage 1 are meant to help promote the Stage 1 client growth goals. The skills are the trees; the goals are the forest. The skills are the means to the goal of client growth.

In Stage 1 the helper skills reflect the helper primarily as role model and catalyst. They include the attitudinal goals of rights, respect, and responsibility, and the behavioral skills of genuineness, positive regard, and boundary distinction. Steps 1–3 of the communication model (listening, centering, and empathizing) are used in Stage 1, together with specific skills for facilitating each step as outlined below (see Part V Table).

Attitude of Rights, Respect, and Responsibility

Attitude overrides what is said. Attitude is based on beliefs and assumptions. Consequently, helper skills will be ineffective unless they are based on an attitude of rights, respect, and responsibility (see Part III).

The behavioral goals of genuineness, positive regard, and boundary distinction are also impacted by attitude. Genuineness with others requires first being able to be congruent. Congruence is based on the belief that you have the right to be yourself. Positive regard *is* an attitude of respect for one's right to be oneself.

In short, attitude affects the ability to use the helper and communication skills effectively. The beliefs and corresponding attitude regarding rights, respect, and responsibility provide a foundation for the helper skills discussed below. Remember, the purpose of the skills is to communicate in a way that facilitates client growth toward a positive self-concept through increased self-understanding. Skills are merely tools. There are many different tools that can work for the same purpose. The ones below are examples. Don't forget the purpose.

Listening

The first step of the communication model is listening. A specific skill that can facilitate the listening process is attending.

Attending

Attending is both the physical process of being alert and ready to receive communication, and the mental and emotional processes of selectively paying attention and being available to concentrate on the communication received. Attending refers to the way you orient yourself physically and psychologically to clients (Egan, 1990).

This skill begins with a mental preparedness to focus on your clients and to understand what they are saying and experiencing.

> [Attending] is accomplished when you actively listen to the client in order to discern the client's primary or essential message; demonstrate an interest, acceptance, and respect for the client and the client's internal frame of reference; and communicate . . . that you understand what the client is saying or is attempting to say. (Doyle, 1992, p. 157)

When one is attending mentally, certain physical behaviors tend to follow. Behaviors such as interested facial expression, body proximity, head nodding, and eye contact typically accompany mental attention. Communication of attentiveness is through facial expression, body messages, and verbal behavior (Hackney & Cormier, 1988). Attending means communicating concern for the client through nonverbal behavior (Ivey & Simek-Downing, 1980).

The purpose of attending is twofold: (1) to prepare yourself to listen, and (2) to communicate to your clients that you are totally present and available to them.

> Effective attending tells clients that you are there with them and that you are ready "to listen." Attending puts the counselor in a position to listen carefully to what clients are saying both verbally and nonverbally. (adapted from Egan, 1990)

As a result of the helper's attending, (1) the client is encouraged to talk, (2) the client is more likely to feel acceptance and positive regard, and (3) the helper can listen and recall more easily.

You can develop and communicate attending behavior in several ways. Hackney and Cormier (1988) suggest communicating attentiveness through three channels:

1. Facial expression. Your facial expressions communicate messages to the client that are as meaningful as those you receive from the client's facial expressions. Three types of facial expressions include eye contact, head nods, and manipulation of facial muscles.
2. Body messages. The body communicates the amount of tension that you are feeling.

3. Verbal behavior. Points to be considered in terms of your verbal impact include these:
   a. Fit your comments or questions into the context of the topic at hand
   b. Don't interrupt clients or jump topics
   c. Stay with the topics that the client introduces and help her or him develop and pursue them

Verderber and Verderber (1989) suggest the following steps for developing attending behavior:

1. Adopt a positive listening attitude
2. Get ready to listen
3. Adjust your hearing to the goals of the situation
4. Make the shift from speaker to listener a complete one
5. Hear a person out before you react
6. Analyze and, if possible, eliminate physical impediments to attending

Egan (1990) uses the acronym SOLER as a formula for attending:

S—Face the client *Squarely*
O—Adopt an *Open* posture
L—*Lean* toward the client
E—Maintain good *Eye* contact
R—Try to be relatively *Relaxed*

Attending is essential to listening. Listening as a communication skill would be much more difficult without the use of attending skills. It is important to remember that attending is mental preparedness linked with corresponding physical behaviors. Our bodies instinctively reflect our underlying beliefs. Consequently, if you have an underlying attitude of acceptance for a person's rights, respect for the person, and encouragement for appropriate responsibility-taking, that will be reflected nonverbally.

Centering

The second step in the communication process is centering. Centering is the process of checking in with yourself in order to become congruent. It has three steps: (1) listening to yourself, (2) being aware of your thoughts, feelings, and behaviors, and (3) differentiating between yourself and others.

Just as you are not in a position to deal with an aggressor physically if you are not balanced, you cannot help a client if you are not first centered.

> Centering involves the helper's being authentic, striving for self-actualization, being in touch with his or her own feelings, and taking responsibility for these feelings. The therapist should be aware of internal and external incongruencies that he or she might be exhibiting in the helping relationship.
>
> The helper's ideal manner of responding is being individual, natural, and unaffected. The therapist should be in touch with his or her own feelings and reactions, and reflect genuinely. (adapted from Corey, 1991)

Centering means acknowledging and integrating the three R's: rights, self-respect, and self-responsibility. It means reminding yourself that you have the right to be yourself, giving yourself permission, therefore, to be congruent and ultimately to experience a feeling of self-acceptance. It includes respecting yourself for your capability and inherent worth, focusing on developing competence and thus experiencing self-esteem. It includes acknowledging that your decisions and choices are your responsibility, resulting in an internalized control and leading to self-actualization.

Centering means that you are acknowledging the right of others to be themselves, respecting their capability and responsibility for making their own choices and decisions relative to their lives. You are acknowledging your own rights, self-respect, and self-responsibility. As a result of these acknowledgments, you are in a position to distinguish between your issues, thoughts, feelings, and behaviors and those of your clients. Three specific skills related to and conducive to centering are genuineness, positive regard, and boundary distinction.

Genuineness

Genuineness, as discussed in Part II, means being one's real, authentic self in the presence of others. Congruence, the internal consistency of thoughts, feelings, and behaviors, is a prerequisite to genuineness.

You can't be yourself with others if you're not able to be and accept yourself. Effective helpers are able to let others see them as they actually are. Reluctance to allow this results from our fears of being rejected.

> We should direct our energy toward accomplishing tasks and solving problems rather than spending it on playing a role or presenting a professional facade. (adapted from George & Cristiani, 1990)

> [Genuineness means that helpers are themselves in the relationship; they avoid presenting a facade or acting with contrivance because they are helpers.] The [helper] must be able to accept all of his own feelings, even those that may be inappropriate to the relationship. (Belkin, 1988, p. 71)

> Rogers states that genuineness is the most important characteristic that a helper can possess. It implies that helpers are real; that is, they are "genuine, integrated, and authentic" in the helping relationship. (Corey, 1991, p. 213)

Positive Regard

Positive regard is the communication of acceptance of your clients as fellow human beings, and respect for their rights to be themselves. For a definition and discussion of positive regard as a concept see Part II.

Positive regard, as it relates to effective listening, is a prerequisite attitude. It encompasses the three R's: rights, respect, and responsibility (see Part III).

Boundary Distinction

Boundary distinction, as discussed in Part II, means being able to distinguish between your personal issues and others' issues. It means knowing where I stop and you begin. It means recognizing the difference between my thoughts, feelings, and behaviors, and your thoughts, feelings, and behaviors.

The purpose of the helping process is to focus on the needs of the client. The ability to distinguish boundaries, to recognize and separate your issues from those of your clients, is an essential skill in being able to focus on the clients' needs. Boundary distinction is a natural outcome of the centering process. Focusing on distinguishing boundaries also facilitates the centering process.

Empathizing

The third step in the communication model, and the last in Stage 1 (see Part V Table), is empathizing. Empathizing is understanding the perspective of your clients, and communicating that understanding to them. Empathizing is being able to look at the situation through your clients' eyes.

> It's important for you to have a sense of the client's private inner world of meaning as if it were your own but without ever losing the "as if" quality. This involves a deep understanding of the client's subjective experience, and feeling with [not for] the client. (Corey, 1991, p. 215)

In empathizing, you can respond to either the verbal message (content) or the nonverbal message (underlying feelings). For example, when your client states "Oh, I'm just pleased as punch to be here today" with sarcasm in her voice, you can respond to the verbal content or the nonverbal message. In this case the two are obviously very different. Two specific skills can facilitate empathizing in Stage 1: explicit empathy—empathizing with what your client has made explicit—and concreteness—clarification of unclear terms and concepts.

Explicit Empathy

Explicit empathy is communicating an understanding of what your client has explicitly stated. Explicit empathy can focus on any one or a combination of the client's basic human dimensions: thoughts, feelings, or behaviors.

For example, your client says:

> "I've just had a horrible fight with my husband. He is just impossible! I might get a little emotional like he says, but he's impossible to talk to—he just clams up. I actually drew my hand back ready to slap him, but stopped myself."

Explicit empathy focused on thoughts:

> "So you think your husband is impossible when he calls you emotional and then is unavailable to talk to."

Explicit empathy focused on feelings:

> "So you felt a lot of emotions during this fight, and it sounds like you still have some very strong feelings now."

Explicit empathy focused on behaviors:

> "So you had a really bad fight with your husband, and actually had your hand pulled back ready to slap him before stopping yourself."

Explicit empathy has been referred to as primary empathy (Patterson & Eisenberg, 1983) and primary accurate empathy (Egan, 1990). It can be described as the ability to respond so that both client and helper have an understanding of the client's main points, perspectives, and experiences.

Empathy should not be confused with paraphrasing. Empathy is an empathic understanding of the client's frame of reference, a communication that the helper understands the client's concerns from the client's point of view. Paraphrasing is simply rephrasing what the client said. Empathy is *understanding* what the client said, from the client's perspective. Paraphrasing may be used to communicate empathy, but it is not necessarily empathizing. In other words, *paraphrasing* is rephrasing what the clients say; *empathy* is *understanding* what the clients feel, think, or experience from their perspective. *Empathic statements* communicate empathy; *paraphrasing* is a tool to use in empathic statements.

Concreteness

Concreteness is "characterized by, or belonging to, immediate experience of actual things or events" (*Webster's New Collegiate Dictionary*, 1981). It means clarifying a word, statement, or concept that is otherwise vague, abstract, or unclear. It means helping your clients speak in terms of the *specific* rather than the *general* when they speak of thoughts, feelings, and behaviors.

For example, if your client says, "My life is a mess," you have no way of knowing what is meant by "a mess." "A mess" could be anything from a broken nail to the loss of a job or death of a partner. Just as with explicit empathy, concreteness can apply to any one or a combination of thoughts, feelings, and behaviors.

Self-analysis of how affectively you are helping your client concretize might include the following questions:

Are you making assumptions about your client's experience?
Are you allowing vagueness or lack of clarity?
Are you trying to guess what your client means by certain words or statements, rather than asking for clarification?
Are you letting your client make generalizations rather than being specific?

HELPING RELATIONSHIP GOALS

The overall process goal for the helping relationship is a growth-conducive atmosphere and environment. The goal for the helping relationship in Stage 1 is the development of rapport and trust (see Part V Table)—specifically, the development of trust through a process of rapport-building. As discussed in Part II, rapport is an important dimension of a helping relationship.

Rapport

Relative to the process framework, rapport is both the result of and can be facilitated by (1) the three R's—rights, respect, and responsibility—and (2) the effective implementation of the Stage 1 helper process skills to facilitate the Stage 1 client process.

Rapport through the Three R's

The successful communication of the three R's—rights, respect, and responsibility—helps promote and develop rapport. Successful communication, however, includes two steps. First, you must really (1) believe that people have the right to be who they are, (2) respect their capability to be the best they can be, and (3) believe they are responsible for their own actions and decisions. Second, you must project an attitude to your clients that successfully communicates that you hold these beliefs.

Rapport through Implementation of the Process Skills

The effective use of the helper process skills in Stage 1 not only helps facilitate the client growth process of telling the story, developing self-awareness, and making the initial problem identification; it also helps promote and develop rapport. In other words, the process skills of listening (attending), centering (genuineness, positive regard, and boundary distinction), and empathizing (explicit empathy and concreteness) are important in making a positive connection with your clients.

Finally, client participation in Stage 1, through telling their story, becoming aware, and identifying their initial problem, contributes to the establishment of rapport. Even when the helper is communicating the three R's and using the Stage 1 process skills effectively, ultimately clients will still choose whether to, and at what level they are ready to, participate in the helping process.

SUMMARY

Stage 1 focuses on the presenting problem and includes the time period from initial contact through initial problem identification.

From the clients' perspective, Stage 1 includes telling of their stories, initial awareness, and initial problem identification. Helper skills, in Stage 1, incorporate the first three steps of the communication model and the corresponding facilitative skills, as shown below.

Communication Step:	Facilitative Skill:
1. Listening	Attending
2. Centering	Genuineness
	Positive Regard
	Boundary Distinction
3. Empathizing	Explicit Empathy
	Concreteness

Relationship goals in Stage 1 are rapport and trust.

18

Stage 2:
Underlying Issues

*Everything that irritates us about others can
lead us to an understanding of ourselves.*

━━ CARL JUNG

tage 2, or the middle stage of the helping process, includes the pe-
S riod of time in the relationship following the establishment of
___ trust and rapport and including processing and understanding
(see Part V Table). Stage 2 is the core of the helping process. Focusing
on the underlying issues—why the presenting problem is an issue for
this particular individual—this stage is the real processing period. Stage
1 lays the foundation, and Stage 2 provides the time for the concentrated and focused work of building the structure.

Just as a good foundation is a prerequisite to building a house, Stage
1 is a prerequisite to Stage 2. The skills of Stage 1 are assimilated into
Stage 2, which lends itself to a wide range of theoretical approaches.

In Stage 2 the client process focuses on the underlying issues. It
includes (1) problem differentiation, (2) examining the issues, and
(3) new perspectives and insight (see Part V Table).

Helper skills in Stage 2 continue to incorporate the attitudinal goals
of the three R's and the first three steps of the communication model
used in Stage 1. In addition, Stage 2 adds step 4 of the communication
model: focusing. This includes both empathic (communicates other-
perspective) and expressive (communicates self-perspective) focusing.
The specific skills of implicit empathy and confrontation are examples

of empathic focusing. Self-disclosure and immediacy are specific skills used in expressive focusing (see Part V Table).

Helping relationship goals for Stage 2 are processing and understanding. The stages are cumulative; therefore, Stage 2 also incorporates rapport and trust.

CLIENT PROCESS

The process for the client in Stage 2 involves moving from initial problem identification (for example, my children don't take enough responsibility for themselves; my friends never call me; my partner doesn't give me enough attention) to a recognition of the issue underlying the presenting problem (such as I feel I'm inadequate as a parent; I feel left out and excluded; I feel unimportant when I'm not receiving attention). The client process during Stage 2 includes (1) differentiating the problem (separating external disappointments from internal underlying issues), (2) examining the issues, and (3) developing new perspectives and insights. These three processes help clarify for clients the issues underlying their presenting problem.

Differentiating the Problem

We all develop external beliefs, which suggest that our well-being is dependent on others and that we are not responsible for our own thoughts, feelings, and behaviors (see Part II). Problem differentiation means separating external and internal issues.

External means everything outside of ourselves (circumstances, other people). External issues are beyond our control; although we may have an influence or impact on them, we cannot direct their outcome. For example, loss of a job, death of a parent, or a divorce action by a partner are all external issues that we may influence and affect but which we cannot control. External issues are generally things we may feel disappointed about but which we need to see as separate from our value or worth. Helping the client with this differentiation is part of the process in Stage 2.

Internal means everything inside ourselves (our thoughts, beliefs, feelings, attitudes, behaviors). Internal *issues* are thoughts, feelings, or behaviors that result in our feeling less than good about ourselves or that contribute to a negative rather than a positive self-concept. They are related to self-concept through internal interpretation. For example, believing you are inadequate if you lose a job, feeling regret for not spending time with an elderly parent before she dies, or concluding you are unlovable because your partner leaves you are all internal issues reflecting internal beliefs, feelings, and actions over which you do have control.

Internal issues have the potential for interfering with the development of a positive self-concept. They reside within the individual and relate to growth goals, to being, doing, and choosing over which the individual has control. Internal issues can affect the client's (1) affective experience goals of self-acceptance, self-esteem, and/or self-actualization, (2) behavioral goals of congruence, competence, and/or internalized control, and (3) cognitive beliefs of rights, self-respect, and self-responsibility.

Internal issues, then, generally cause clients to feel incongruent or to experience discrepancies between beliefs and actions, to perceive themselves as inadequate or incompetent, and to yield to an externalized locus of control. These issues involve how we feel about ourselves; therefore, there is something we can do about them.

Problem differentiation involves, first, distinguishing between the external and internal, and, second, shifting the focus to the internal issues. You must determine what is your problem and what isn't, and take responsibility for what you can change.

> As soon as clients begin focusing on their own behavior, they will start to feel less "stuck," powerless, and depressed. Most clients will start to see more alternatives available to them and begin to feel more hopeful and in charge of their lives. Thus, the therapist's task is to discourage clients from passively complaining about others or from trying to manipulate and control others' behavior. Although clients usually fail in their attempts to change others, they can often succeed in resolving problems by gaining greater mastery over their own responses. (Teyber, 1992, p. 74)

The client focuses more and more on himself rather than on others—moving to a rich, flowing, changing, authentic creative self. This person trusts his own experience and life processes. He uses his own feelings ("guts") as a point of departure for understanding the world. He will consult his own feelings as the ultimate guide when confronted with a problem. (Løvlie, 1982, p. 78)

The first component of the internal focus was to help clients look within and become more aware of their own reactions and responses. In addition to looking within, the client must also begin to act from within, by adopting an internal locus for change. (Teyber, 1992, p. 77)

Encouraging clients to make owning, or "I-statements" (see Part IV), will facilitate the shift from an external to an internal focus. I-statements can be used with thoughts, feelings, and behaviors. The following are examples of nonowning and owning client statements, focused on thoughts, feelings, and behaviors:

OWNING A THOUGHT:
Client's non–I-statement: "I want to know what you think about women serving in combat roles."
Client's I-statement: "I think women should serve in combat roles."

OWNING A FEELING:
Client's non–I-statement: "He is impossible when he behaves like that."
Client's I-statement: "I feel hurt and frustrated at his behavior."

OWNING AN ACTION:
Client's non–I-statement: "The car crashed into the garage door."
Client's I-statement: "I crashed the car into the garage door."

The following are additional ways to encourage clients to use owning statements:

1. *Respond as though clients used I-statements.* You can respond to clients in ways that use the word *you* as though they had used an I-statement, even when they have not. For instance, to a client who says, "He is impossible when he behaves like that," you might respond, "You feel hurt and frustrated at his behavior." Your response implicitly encourages the client to express feelings directly.

2. *Request the use of the first person singular.* If clients repeatedly fail to use I-statements, you might consider openly drawing this to their attention. However, you have to judge whether this intervention will be threatening to them. Also, you may have the further decision of how to handle clients' behavior if they revert to their old ways.

3. *Model using I-statements.* If you are open in your own behavior and use I-statements to own your feelings, thoughts, and actions, your example may help clients to do likewise. (Nelson-Jones, 1993, p. 149)

Examining the Issues

Internal issues are those that relate to the self—how clients feel about themselves and how they are living their lives. Internal issues, then, relate to one's self-concept. Self-concept, as discussed in Part II, includes three dimensions that are tied to our basic human dimensions: self-acceptance (being), self-esteem (doing), and self-actualization (choosing). These dimensions reflect how people feel about themselves (being) and how they feel about the way they are living their lives (choosing/doing). The dimensions are reflected in our congruence, competence, and internal control.

The reason to have clients examine the issues is to help them

1. Clarify and further differentiate the underlying issues that are interfering with a positive self-concept
2. Decide what they need to do differently to feel better about themselves and their lives
3. Do it

One approach is to have them examine how they view themselves relative to these three basic components and dimensions: being, choosing, and doing (see Part V Table).

Being: Rights, Congruence, and Self-Acceptance

The dimension of being is the affective, subjective dimension of how you *feel* about yourself based not on what you *do* but on how you *are*. The dimension of being includes your feelings and your belief in the right to be who you are. When you believe you have that right, you have permission to be congruent with your thoughts, feelings, and behavior, and to develop an integrated identity. Belief in the right to be allows for congruence and results in a feeling of self-acceptance. Self-acceptance comes, not from what we do, but from accepting our being—who we are.

> You are worthy because you are. (John-Roger & McWilliams, 1991, p. 357)

> Unhappy people are not in tune with themselves. (Gladding, 1992, p. 65)

Doing: Respect, Competence, and Self-Esteem

The dimension of doing involves *acting*. It focuses on how you feel about yourself based on your actions. Doing includes your behaviors and your self-respect for your capabilities. It includes the need to feel and develop competence, and results in a sense of self-esteem. Self-esteem is the ability to gain desired outcomes as a result of one's own behavior (Craig, 1989). Happiness is learning to enjoy the process of doing and living, regardless of the outcome.

Choosing: Responsibility, Internal Control, and Self-Actualization

The dimension of choosing involves *thinking*, deciding, and making choices. It is choosing how to live your life. This dimension focuses on decisions and life choices that only you can make in order to be true to yourself. The dimension of choosing involves taking responsibility for your actions and life's choices. It involves making choices based on self-approval or internalized control and results in an actualizing of the self. Happiness is attained through changes in behavior, resulting in options and choices that allow greater freedom.

Developing New Perspectives and Insights

New perspective means seeing the situation in a new way—from a different angle. Insight means understanding the inner nature of things (*Webster's New Collegiate Dictionary*, 1981).

In facilitating client growth, new perspectives and insight involve two of the three goals for clients' examining the issues: (1) clarifying what the issue is and (2) seeing what they need to do differently to feel better about themselves. New perspectives and insights involve shifting the focus from the external to the internal. This requires a new understanding of how clients feel about themselves and/or how they are living their lives. This can include a new understanding of any one or a combination of self-concept components: self-acceptance, self-esteem, or self-actualization.

HELPER SKILLS

The helper skills in Stage 2 build on and incorporate those from Stage 1. The purpose in Stage 2 is to help the client clarify what the real or underlying issues are. This is done by facilitating the client growth goals of (1) differentiating and shifting from external to internal issues, (2) examining the internal issues, and (3) promoting new perspectives and insights.

In Stage 2 the helper is no longer just a role model and catalyst but becomes an active facilitator, purposefully intervening to promote these client growth goals. Stage 2 incorporates, in addition to the first three steps, step 4 of the communication model: focusing (see Part V Table).

Focusing

Focusing is central to the helping process. It is the intentional directing by the helper of attention toward specific aspects of client communication for the purpose of facilitating clarification, insight, and understanding. It refers to both (1) the process of directing attention to and concentrating on significant points, and (2) the resulting clarification of an image, concept, or insight.

The purpose of focusing is to facilitate client self-understanding through differentiating internal from external issues, examining the underlying internal issues, and gaining new perspectives and insights. While focusing on explicit content takes place as a result of explicit empathy and concretizing, the skill of "focusing" itself refers to intentional focus beyond primary level, explicit content, to implicit, underlying issues.

In the process, you can focus on (1) one or more of the human dimensions—cognition (thoughts, choosing), affective experience (feeling, being), or behaviors (actions, doing)—and (2) verbal and nonverbal behavior. Verbal and nonverbal behavior include both explicit and implicit messages. The explicit message of verbal behavior is the content—"You look great"—while the implicit message is the underlying intent of the content—jealousy over how great you look. The explicit message of nonverbal behavior is the behavior—slamming a book down on the table—while the implicit message is the underlying content of the behavior—anger at information in the book.

To summarize, focusing can be directed at three categories: human dimensions, verbal behavior, and nonverbal behavior:

1. Human Dimensions
   • cognition (thoughts) (explicit/implicit)
   • affective experience (feelings) (explicit/implicit)
   • behaviors (actions) (explicit/implicit)
2. Verbal Behavior
   • explicit (content)
   • implicit (underlying message of content)
3. Nonverbal Behavior
   • explicit (behavior)
   • implicit (underlying message of behavior)

Focusing can be accomplished through both empathic and expressive focusing. Empathic focusing uses the clients' perspective; expressive focusing uses the helper's perspective. Following are two possible responses to a client statement. Both address client termination. However, the focus of one is empathic (client perspective); and the focus of the other is expressive (helper perspective).

Client: I think I'd like to try cutting back to coming in once every two or three weeks for awhile. Things have been going pretty well, and I think I can manage coming in less often.

Helper: (empathic focus) It sounds as though, since things have been going so smoothly, you're starting to think about how to bring our sessions to a close.

Helper: (expressive focus) I've been noticing that things seem to be going more smoothly for you and that you've brought up fewer unresolved issues. Our sessions haven't seemed to have quite the relevance lately, either. It feels to me, too, like this might be a good time to talk about cutting back, and even when you would like to terminate our sessions.

The focus may be the same, whether the approach is empathic or expressive. The difference is whether it is viewed from the client's or the helper's perspective.

Empathic Focusing

Empathic focusing means (1) directing attention to clients' thoughts, feelings, and/or behaviors by means of their verbal and/or nonverbal communication through (2) communicating an understanding of the *client's* perspective. A major effect on clients of being empathically understood is a focusing effect. Empathic attunement stimulates those who perceive they are being understood in a way to further attend to their own experiencing. The emphasis on the client is key to the focusing effect. Empathic understanding responses tend to facilitate more communication from clients concerning themselves. Clients tend to become more consistently and intently focused (Corey, 1991).

Two specific helper skills that reflect empathic focusing are implicit empathy and confrontation.

Implicit empathy   Explicit empathy, used in Stage 1, is communicating an understanding of what the client has *explicitly* stated or demonstrated, verbally or nonverbally, regarding thoughts, feelings, or behaviors. Implicit empathy can also focus on the client's thoughts, feelings, and/or behaviors and either verbal or nonverbal communication. Implicit empathy, however, focuses on the *implicit* message of verbal behavior (underlying message of the behavior). Implicit empathy is communicating an understanding of what the client has only implied. In other words, the clients themselves may be unclear, vague, or confused about their thoughts, feelings, or behaviors. Implicit empathy is looking through the clients' eyes and helping them clarify their own perspectives. Implicit empathy has been described in various ways:

... sensing meanings of which a client is scarcely aware. (Rogers, 1980, p. 142)

Advanced empathy is a process of helping clients explore ideas, thoughts, and feelings new to their awareness. (adapted from Patterson & Eisenberg, 1983)

"Advanced Accurate Empathy"—communicates an understanding of what the client only implies, or has poorly formulated. (Egan, 1994, p. 214)

Implicit empathy is, in a sense, understanding your clients' perspectives before they do. Keep in mind, however, that implicit empathy is focusing on what the client is confusedly saying, but *is* saying or implying, however vaguely. It is *not* the helper's *interpretation* of what the client is saying. Implicit empathy can focus on thoughts, feelings, or behaviors:

Client: I just can't believe my daughter's going to marry an unemployed, uneducated atheist! With all her potential, and after all we've done for her to help her get an education. I can't believe she'd do this to me!

Helper: It sounds as though all that time that you were helping your daughter, particularly in getting an education, it just didn't occur to you that she might choose to do something like this—something that would be disagreeable to you. (implicit empathy focused on thought)

Helper: It almost sounds as though your daughter's actions feel like a personal affront to you. Like you feel she's doing it purposely to hurt you. (implicit empathy focused on feelings)

Helper: So you've done an awful lot for your daughter, particularly in helping her through school, and it almost sounds as though, if you'd known this would happen, you might have reassessed what you were doing for her. (implicit empathy focused on behavior)

Several potential problem areas can interfere with the effective use of implicit empathy. These are illustrated below with alternative responses that might have been made to the client statement above.

1. Imposing helper perspective (i.e., interpreting): "It seems to me, you're just having a hard time letting go of your daughter. You don't want her to have a life of her own!"
2. Inaccurate empathy: "It sounds as though you think she's being coerced into this marriage."
3. Judgmentalness: "Don't you want your daughter to make her own choices and be happy?"

Confrontation Confrontation, in this model, is a form of empathic focusing and a specific type of implicit empathy. Contrary to the typical view as a clashing of ideas or forces with one person confronting another, confrontation here means that the helper *empathizes* to help clarify the clients' perspectives. In a case where confrontation would be used, the clients' perspectives reflect contradictions or discrepancies of which the client is unaware. It is through the process of empathic focusing on these contradictions and discrepancies that the client clarifies underlying issues. Confrontation, then, is empathically focusing on contradictions and discrepancies from the client's perspective.

Confrontation is

. . . a verbal response in which the counselor describes discrepancies, conflicts, and mixed messages apparent in the client's feelings, thoughts, and actions. (Cormier & Cormier, 1991, p. 116)

. . . pointing out incongruity, discrepancies, or mixed messages in behavior, thought, and feelings. (Ivey, 1991, p. 90)

. . . challenging discrepancies in clients' lives and inviting explorations of discrepancies, distortions, games and evasions, without being judgmental or non-supportive. (Egan, 1994, p. 234)

The following client-helper exchange illustrates empathic confrontation:

Client: I really want my son to be his own person, to think for himself, and make up his own mind. But, in a case like this, where his future's at stake, he better decide to go to college!

Helper: So on the one hand, you really want your son to make up his own mind, and on the other hand, it's really important—at least in terms of some decisions—that he make a decision that would be the one you would make.

Confrontation invites clients to look at discrepancies in their thinking, feelings, or behaviors, and reassess their perspective. As a result, confrontation helps them gain greater clarification of underlying issues. Discrepancies can occur in (1) verbal and nonverbal behavior, (2) thoughts, feelings, and behaviors, and/or (3) perceived reality and actual reality.

In verbal and nonverbal behavior, these might include discrepancies

1. between two statements. For example, your client says at one
   point she cares very much about her partner, and later says she
   hates her partner.
2. between two behaviors. For example, your client may be smiling
   while crying.
3. between a statement and a behavior. For example, your client
   may *say* she's angry, while smiling. Another may say he's com-
   mitted to counseling, while continuing to miss appointments.

Discrepancies in thoughts, feelings, and behaviors might occur

1. between stated thoughts and feelings: For example, your client
   may say she believes married partners should stay together but
   that she feels suffocated and unhappy.
2. between stated thoughts and behavior: Your client says she believes marriages should last forever, but she has left her partner.
3. between stated feelings and behavior. Your client says he feels
   excited about the promotion but has a totally flat affect and is
   listless.

Discrepancies in perceived reality and actual reality might include

1. discrepancies in observed reality. For example, your client may state she's worthless, ugly, and stupid, while you can observe that in reality she's worthy, attractive, and intelligent. She may state "I have no friends" while you observe that someone came with her and is waiting for her in the waiting room.
2. polarized thinking. Your client goes from "everything's perfect" to "everything's a mess."
3. failure to acknowledge choices. Your client states, "My children *have* to go to private school," without acknowledging there is a choice.

Effective confrontation is empathic focusing on discrepancies to help the client gain self-understanding through clarifying underlying issues. Confrontation should reflect the goal of the helping relationship: to help clients understand themselves better. Effective confrontation includes the following checklist:

1. ATTITUDE OF THREE R'S
- Rights—clients have a right to be where they are, including having defenses, and choice as to when they are ready to move. Honor defenses and don't push.
- Respect—respect your clients' right to make their own choices. Don't judge or patronize.
- Responsibility—ultimate responsibility for choice and change rests with your clients. Let them have that responsibility. Avoid giving advice.

2. BEHAVIORAL CHECK
- Genuineness—be genuine and honest.
- Positive regard—communicate verbally and nonverbally an acceptance of your client.
- Empathic focusing—communicate empathy; sincerely try to understand the *client's* perspective rather than presenting the *helper's* perspective.

3. RESULTING AFFECTIVE EXPERIENCE
- Empathy—communicate to clients that the helper is empathizing with *their* dilemma, their experience.

- Tentativeness——communicate to clients that the helper is suggesting a possibility for them to consider and assess, not telling them how they think, feel, or should behave.

Ineffective confrontation does not result in an empathic focus on discrepancies from which the client gains self-understanding through clarifying underlying issues. When confrontation is ineffective, it's generally a result of the following common potential problem areas, listed with examples:

1. Judgmentalness: "It sounds like you just want things your way."
2. Not acknowledging discrepancies: "So you really want the marriage to work" (when client also said she left her partner).
3. Failure to empathize: "You need to get clear about what you want!"
4. Imposing conclusions or assumptions: "It sounds like you need to apologize to your partner."

Expressive Focusing

Expressive focusing means (1) directing attention to clients' thoughts, feelings, and/or behaviors by means of their verbal and/or nonverbal communication by (2) communicating an understanding of the *helper's* perspective. Notice that even though expressive focusing uses the helper's perspective, the intent is still to direct attention to the *clients'* thoughts, feelings, and behavior. In using expressive focusing, the purpose is still the same as in empathic focusing: to help the client gain self-understanding through clarifying underlying issues. Expression is one of the core skills in helping relationships. If the helper cannot respond after having listened and attended, then that helper will not be able to help clients achieve their goals. Expressing involves more than the response skill of empathy (Doyle, 1992). Two specific helper skills that reflect expressive focusing are self-disclosure and immediacy.

Self-disclosure   Self-disclosure, literally, is the disclosing of one's self to another. A certain amount of indirect self-disclosure automatically happens through interaction. In a helping relationship, as the helper, you disclose yourself indirectly through your congruence, openness, and genuineness. By *being* yourself, you are indirectly disclosing characteristics about yourself.

The [helper] communicates his or her self to the client in every look, movement, emotional response, and sound, as well as with every word. Clients actively construe the personal characteristics, meanings, and causes behind the [helper's] behaviors. (Strong & Claiborn, 1982, p. 173)

Self-disclosure, as a helper skill in expressive focusing, is direct and purposeful. It refers to the verbal disclosure by counselors of their own thoughts, feelings, and/or behaviors related to their experience for the purpose of focusing on and facilitating client self-understanding.

Self-disclosure means a sharing of personal information about yourself, your experiences, your attitudes, and your feelings. (Evans, Hearn, Uhlman, & Ivey, 1989, p. 158)

Self-disclosure should be used appropriately and not indiscriminately in the counseling sessions. (Hackney & Cormier, 1988, p. 22)

Effective self-disclosure is used to focus on and facilitate client self-understanding. It is to benefit the client in examining and clarifying underlying issues. It does not distract or take the focus off the client and the goal of the helping relationship. The following illustrates effective self-disclosure:

Client: I am so overwhelmed. Graduate school is really demanding. Between trying to keep up with assignments, working part-time, and raising my daughter, I have no time for myself. I'm really on overload.

Helper: When I was in graduate school it seemed there was always a demand or deadline. I sometimes felt that somebody else actually had control of my life—I felt out of control. I'm wondering if that's at all like what you're experiencing.

The self-disclosure relates to what the client is saying and invites the client to examine her experience from a different perspective. The helper closes the self-disclosure by bringing it directly back to the client.

Ineffective self-disclosure, on the other hand, takes the focus off the client and away from the goal of facilitating client self-understanding. It distracts or overwhelms the client. For example, an ineffective self-disclosure response to the previous client statement might be this

> *Helper:* I know what you mean. When I was in graduate school, there were so many demands and assignments I could hardly keep up. In fact, I failed a couple of courses, and almost didn't make it through.

In this case, the self-disclosure is distracting and potentially overwhelming. It takes the focus off the client and there's no attempt to relate it back to the client's issue.

The purpose of self-disclosure, like the other skills in Stage 2, is to facilitate client self-understanding through examining underlying issues and developing new perspectives and insights. Self-disclosure provides a nonthreatening invitation to the client to consider another perspective. It has several advantages:

1. Effective role modeling—particularly for clients trying to develop skills in self-expression
2. Validation or universality of feelings—especially in cases where clients believe that certain feelings are unacceptable
3. Permission-giving—probably the most nonthreatening invitation to consider reassessing one's perspective, such as, "If I were you, I think I might feel angry"

On the other hand, self-disclosure can be used negatively:

1. Distracting or overwhelming the client—self-disclosure must be carefully used to prevent distracting your clients from *their* issues.
2. One-upping your client—having the self-disclosure seem more important or significant than the client's issues.
3. Role-reversal—having the client shift focus to trying to help the helper.

In using self-disclosure, the helper has a number of choices: (1) whether to disclose, (2) how much to disclose, (3) what aspects to disclose (e.g., just information about an experience, or feelings about the experience), (4) effect of the experience (e.g., how you dealt with it, what you learned from it), and (5) current perspective and/or feelings about it. The following guidelines can help regarding use of self-disclosure:

1. When in doubt, don't use self-disclosure. If there's any question of whether it would be effective, use something else.
2. Be very selective. Self-disclosure is the one skill you might never use and still be effective. Therefore, if you wonder whether it would be irrelevant, overwhelming, distracting, or just not helpful, don't use it. There is always another intervention you can use instead of self-disclosure.
3. Be brief. Don't take the focus off the client and the client's issue.
4. Refocus attention back to the client. The purpose of self-disclosure is to provide an invitation for the client to reassess his or her perspective. Provide the connection between your self-disclosure and its relevance to what the client is experiencing.
5. Observe client reaction. If disclosure seems to be having a negative effect, check it out and be prepared to deal with it with immediacy (see below).

Potential problem areas include these:

1. Distracting or overwhelming self-disclosure:

> *Client:* I'm really upset about how my parents' visit ended. We had a big fight just before they left and we never got it resolved.
>
> *Helper:* I can really relate to that. I had a big fight with my father the last time he visited. He left with it unresolved, and then he died before I got to talk with him again.

2. Irrelevant disclosure:

> *Helper:* Yeah, I've always had trouble even relating to my parents. Not that I don't love them, but we're totally different.

3. Role-reversal and focusing on the helper:

> *Client:* Boy, it's too bad about your father's death. How did you deal with having a lack of resolution before he died? Did you feel guilty?

Immediacy   Most client issues, either primarily or secondarily, relate to problems with interpersonal relationships. Since the helping relationship is a type of interpersonal relationship, to some extent it will be parallel to other relationships in the client's life. It will mirror the issues that the client struggles with in these relationships. For example, a client who is compliant in interpersonal relationships will undoubtedly be compliant at some point in the helping relationship. Consequently, the helping relationship provides a powerful opportunity to address general relationship dynamics with the client.

Immediacy means the quality or state of being immediate (*Webster's New Collegiate Dictionary*, 1981). In a helping relationship, immediate means happening in the present, or within the helping relationship. The skill of immediacy is *addressing* relationship dynamics within the context of the helping relationship.

> Immediacy involves an understanding of what is going on between the helper and client within the helping relationship. (adapted from Gladding, 1988)

> Immediacy is an exploration of the counselor/client relationship. (adapted from Carkhuff, 1969)

> It helps the client to focus on the here-and-now, including the helper/client relationship. (adapted from Kottler & Brown, 1992)

> Addressing what is happening in the here-and-now between helper and client is a way of exploring the client's interpersonal style and seeing the self in alternative frames of reference. (adapted from Egan, 1990)

There are two types of immediacy: long-term dynamic immediacy and current dynamic immediacy. Both focus on dynamics within the relationship as they relate to the client's issues.

Long-term dynamics is a focus on the overall relationship—its development, direction, tone, and general climate. For example, "You've talked about how, when you date, you never know what to talk about or where to start. I just wanted to share with you that over the last few months, I've noticed I generally ask where you want to start, and you usually have something in mind. I'm wondering if there's something about our relationship that feels different compared to the ones you've spoken of."

Current dynamics is a focus on what's happening right now in the relationship. For example, "Just now when I asked where you wanted to start, you didn't know. I'm wondering if there's something about how I asked the question that affected you."

You'll notice that immediacy incorporates several other skills we've already discussed. It may contain elements of self-disclosure, empathy, and/or confrontation. For instance, in the long-term dynamics example above, you might identify the following:

- Self-disclosure—"I just wanted to share that over the last few months, I've noticed I generally ask where you want to start . . ."
- Empathy—"You've talked about how, when you date, you never know what to talk about or where to start."
- Confrontation—"I'm wondering if there's something about our relationship that feels different compared to the ones you've spoken of." (addressing implied discrepancy)

The purpose of using immediacy is to promote client self-understanding by examining underlying issues and gaining new perspective and insight. Specifically, immediacy invites the client to examine interpersonal dynamics within the context of a safe, accepting, and nonthreatening environment. As with self-disclosure, it's important to relate the dynamic back to the context of the client issue.

Potential problem areas in use of immediacy include these:

1. Ignoring relationship dynamics: An illustration would be for the helper in the example above to focus only on the client's dating relationships.
2. Judgmentalness: "I've noticed that in our relationship you never seem to have a problem knowing what to talk about; why can't you just do that in your dating relationships?"
3. Interpreting or drawing conclusions: "I'll bet I remind you of your father, and that's why you feel more comfortable talking in this relationship; am I right?"

HELPING RELATIONSHIP GOALS

The goal for the helping relationship in Stage 1 was to develop rapport and trust. In Stage 2, the goal is to incorporate that rapport and to provide a safe, supportive environment for the effective processing of client

underlying issues. The process goal is effective processing. The outcome goal is understanding of client underlying issues (see Part V Table).

Processing

Processing is the communication and interaction that takes place between helper and client in the helping relationship. Effective processing is focused on the client issues to facilitate client self-understanding. Effective processing is facilitated by (1) the three R's (rights, respect, and responsibility), and (2) effective implementation of the Stage 1 and 2 helper process skills to facilitate the Stage 1 and 2 client process.

Processing through the Three R's

The helper's beliefs and attitudes regarding client rights, self-respect, and self-responsibility lay the foundation for effective processing (see Part III).

Processing through Implementation of the Process Skills

The cumulative helper communication and process skills in Stage 1 and Stage 2 include listening (attending), centering (genuineness, positive regard, and boundary distinction), empathizing (explicit empathy, concretizing), empathic focusing (implicit empathy, confrontation), and expressive focusing (self-disclosure, immediacy). These skills are used to promote processing. Their effective use should facilitate client participation in client growth goals. Client growth goals in Stage 2 focus on underlying issues, including differentiating the problem, examining the issues, and developing new perspectives and insights.

Understanding

The outcome goal of the helping relationship in Stage 2 is understanding the clients' underlying issues (see Part V Table).

SUMMARY

Stage 2 focuses on the underlying issues (of the presenting problem) and includes the period of time from initial problem identification through the development of new perspectives and insights. From the clients' perspective, Stage 2 includes differentiating the problem, examining the issues, and developing new perspectives and insights.

Helper skills in Stage 2 incorporate the fourth step of the communication model and the following corresponding facilitative skills.

Communication Step:	Facilitative Skills:
4. Focusing	
Empathic Focusing	Implicit Empathy
	Confrontation
Expressive Focusing	Self-Disclosure
	Immediacy

Relationship goals in Stage 2 include processing and understanding.

19

Stage 3:
Direction and Change

*Don't be afraid to take a big step
if one is indicated. You can't cross
a chasm in two small jumps.*

— DAVID LLOYD GEORGE

tage 3 is the final part of the helping process. It follows processing
and understanding of the underlying issues and incorporates direction and change. The helper has effectively helped the client
focus on underlying issues, differentiate external from internal issues,
examine them, and gain new perspective and insight. In Stage 3 the client is ready to implement the new perspectives in movement toward
and ultimate achievement of a desired change (see Part V Table).

To use the building construction analogy, if Stage 1 lays the foundation, and Stage 2 builds the structure, Stage 3 caps it off with a roof.
Stage 3 is the closure, the implementing, the doing.

Stage 3 is cumulative. It assimilates the skills and goals from Stages
1 and 2. It also lends itself more to a behavioral approach than the
other stages, as it focuses more on actual change in behavior.

In Stage 3, the client process focuses on direction and change. It includes (1) determining direction, (2) implementing a strategy for
change, and (3) encouraging actual behavioral change in the client.

Helper skills continue to incorporate the attitudinal goals of the
three R's and the first four steps of the communication model used in
Stages 1 and 2. In addition, Stage 3 adds step 5 of the communication

model: directional support. Specific skills used in directional support include directionality and implementational support. Helping relationship goals for Stage 3 are directionality and change (see Part V Table).

CLIENT PROCESS

The process for the client in Stage 3 involves moving from the new perspective and insight gained in Stage 2 ("I feel unimportant when I'm not receiving attention") to making a decision and doing something about it through the implementation of change in Stage 3 (changing my belief that importance is based on attention, and/or finding a new situation where I get more attention). The client process in Stage 3 includes (1) determining direction, (2) implementing a strategy for change, and (3) changing. These three processes are meant to help the client develop direction and bring about desired change.

Direction Determination

Determining direction means making a decision about which way to go and what to do. This choice generally emerges and becomes self-evident as a result of the new perspective and insight gained in Stage 2. It typically includes two aspects: (1) release of external focus, and (2) acceptance of responsibility for internal issues.

Release of external focus involves letting go of the belief that happiness or a positive self-concept is dependent on external forces. External forces (others' approval, getting a promotion, getting a particular job) do not control whether you have a positive self-concept; and you do not have to control external forces (win others' approval, get your partner to listen to you, have your kids at the top of their class) in order to have a positive self-concept. Release of external focus generally involves letting go of or discontinuing something you've been doing. Examples might be (1) affectively, no longer worrying about what others think of you; (2) cognitively, discontinuing to believe that your worth is based on others' approval; or (3) behaviorally, no longer trying to get your partner to act in a particular way.

Acceptance of responsibility for internal issues involves recognizing that our happiness and a positive self-concept are internal. *Because* they are internal, *only* we can take responsibility for them. The development of a positive self-concept is *dependent* on our accepting responsibility for it. Accepting responsibility means determining what we need to feel good about ourselves and our lives, and then *doing* it. Consequently, while release of external focus means letting go or no longer doing something, acceptance of responsibility for internal issues means taking on responsibility for doing something else. In short, a decision is made to stop going in one direction (external focus) and to start going in another direction (internal focus).

> In the beginning stage of therapy, most clients see the source of their problems in others. . . . Therapy begins with these externalizing complaints. (Teyber, 1992, p. 69)

> Clients who can focus on themselves and see their own participation in a conflict are usually motivated to change their own part in it. (Teyber, 1992, p. 73)

Implementation Strategy

Once the new perspective and insight are clear and the decision made to go in their direction, the next step is devising a strategy to implement the decision. As Albert Einstein said, 50% of any solution is clarifying the problem; however, that still leaves 50% to reckon with. We all know from personal experience that just because we decide we want to be different or do something differently (e.g., stop losing our temper, stop compulsive eating) the change doesn't automatically happen. We have to develop a plan and work at implementing the change of behavior. Implementation means "the bringing about of an alteration, a thing accomplished, usually over a period of time, in stages, or with the possibility of repetition" (*Webster's New Collegiate Dictionary*, 1981).

Change

After the client makes the decision to go in the direction that emerged from the new perspective and insight, through an identified implementation strategy, the final step is the actual behavioral change. This change can be overt or covert. It can be a change in one or a combination of the human dimensions: feelings (being), behaviors (doing), and

thoughts (choosing). It might involve greater acceptance of oneself (feelings/being). It might be becoming more assertive in a relationship (behavior). It might be a change in the client's belief (thoughts) system, such as "no longer being dependent on what others think of me but relying instead on my own approval of myself."

Change simply means "to make different" (*Webster's New Collegiate Dictionary*, 1981). The reality is that unless some kind of actual change occurs, the helping relationship has not been effective.

> [Outcome goals are] directly related to your client's changes, to be made as a result of [helping]. (Hackney & Cormier, 1988, p. 110)

> There is nothing magic about change; it is work. If clients do not act in their own behalf, nothing happens. (Egan, 1990, p. 37)

HELPER SKILLS

The helper skills in Stage 3 build on and incorporate those from Stages 1 and 2. The purpose in Stage 3 is to facilitate client direction and change. This is done by facilitating the client growth goals of (1) establishing direction, (2) developing a strategy for implementation of change, and (3) changing. In Stage 3 the helper is a role model, catalyst, and facilitator. In addition to the continuation of steps 1–4 of the communication model and their respective specific skills, Stage 3 adds step 5 of the communication model: directional support. Specific helper skills used in directional support are directionality and implementational support.

Directional Support

Directional support is follow-through. Just as a hitter's follow-through adds speed, distance, and stability to a baseball, directional support adds power and momentum to the direction the client is going. To provide directional support, two specific skills can be useful to the helper: directionality and implementational support.

Directionality

The purpose of the helping relationship is to facilitate client self-understanding. This involves helping clients explore and examine underlying issues that are interfering with their ability to feel positive about themselves and/or their lives in some way. That's why they have come for counseling. Your purpose is to help them better understand their issues, gain new insights, decide what they need to do, and make changes in themselves and their lives. Implicit in this process is movement. As one goes through the process of moving from initial problem identification, to problem differentiation, to examining the underlying issues and gaining new perspective and insight, there is direction that *emerges* from this process. This direction is not imposed on the process at the outset; rather, it emerges as a *result* of the process. This emergence of direction is what we refer to as directionality.

Directionality, as a skill, means facilitating the emergence of direction. It does not mean *determining* the direction or making decisions for your client. It means facilitating the *emergence* of direction.

When directionality does not seem to be emerging, the following may help:

1. Problem differentiation: Has the client clearly identified the problem? This requires moving from initial problem identification (external focus: "My boss is insensitive") to problem differentiation or recognition of why the issue is a problem for the client (internal focus: "I feel inadequate when he criticizes me").

2. Recurrent themes: Are there recurrent themes? Often these emerge as underlying issues. For example, a client talks about feeling angry at work when an authoritative boss comes on board; later he speaks of anxiety over his wife's going back to work; then he tells that his son is "defying" his desire for the son to attend his alma mater. Recurrent theme: possibly feeling out of control?

3. New insights: Is the client gaining new perspectives and insights? Insight and new understanding are key to directionality.

The following is an example of a client/helper interaction when directionality is emerging:

Client: I don't know why, but I'm feeling just awful about my mother's going into surgery. I know it can prolong her life, but at 85, I'm not sure that's what's best.

Helper: So, even though you know this surgery has the potential of prolonging your mother's life, it sounds like you're not sure that prolonging her life is what's most important.

Client: Exactly. I think she's going in because my brothers have convinced her she should. I'm not sure she, herself, wants the surgery.

Helper: So you're wondering if she's having the surgery just to please your brothers.

Client: Yeah. And here she is, one more time, just doing what she's being told to do. That's the story of her life. And I'm going along with it.

Helper: It almost sounds as though you're relating to your mother's position. It sounds like you're not only feeling bad about your mother's going into surgery, but recognizing there's a theme in her life of doing what she's told; you're empathizing with going along with whatever you're told to do.

Client: (sigh) That's right. I'm just like her. We both have lived our lives doing what we're told. I'm sick of it.

Directionality emerges as the client moves from an awareness of feeling bad about her mother's going into surgery, to realizing that her mother has probably agreed to the surgery to please others, to an insight that she's doing the same thing (doing what she's told) and is tired of it. Directionality results in greater and greater clarification of the issue, an understanding of it, and ultimately, a change in behavior.

The purpose of directionality, as a skill, is to monitor the effectiveness of the helping. If it seems that you are going nowhere, then maybe you're not. If that's the case, you need to address it.

Directionality is addressed as a skill in Stage 3 because it becomes most apparent in this stage, as one views the movement through the stages. The roots of directionality, however, begin in Stage 1 and continue to grow through Stages 2 and 3. There should be a sense of movement and direction surfacing earlier, even though the blossoming occurs in Stage 3.

Directionality can emerge whether the helper is using an indirect (client-centered) or a more direct (rational-emotive therapy) approach. It cannot be imposed by the helper; the helper's role is to assist clients to recognize for themselves the direction they need to take to feel better about themselves and their lives. The client alone is in a position to recognize this.

> [Helping] occurs when the counselor follows the client, rather than when the client is following the counselor. The responsibility for the content and sequence of the process is with the client. Since one of the desirable outcomes of therapy is for the client to assume responsibility for himself, the therapy process should give him that responsibility from the beginning. (Patterson, 1983, p. 23)

Several potential problem areas can interfere with directionality:

1. Wheel spinning—staying focused on concern about her mother's going into surgery when there are underlying issues.
2. Lack of direction—going from concern about her mother's surgery to talking about other irrelevant surgeries.
3. Imposition of irrelevant helper perspective—helper's talking about improvements in medical technology and subsequent reassurance regarding successful surgeries.

Implementational Support

Implementational support means assisting the client in developing and applying a strategy to implement the identified desired change. A prerequisite to implementational support is clear identification of the goal and desired change, which comes through directionality. The purpose is to help the client make the transition from a cognitive decision (e.g., desiring to be more assertive) to the implementation of a change in behavior (e.g., *being* more assertive).

> It is your role to hear the unmet needs as clients describe their problems, to help them hear those needs, to help clients translate these needs into wants, to help them formulate goals that will help meet the needs, and to help develop plans of action. (adapted from Hackney & Cormier, 1988)

> Help clients formulate strategies and plans to accomplish goals and get the new scenario on line. Help them implement their plans. (Egan, 1994, p. 286).

Implementational support involves three elements: strategy identification, strategy implementation, and strategy assessment.

Strategy Identification

After the client's goal and desired change are clarified, a strategy needs to be identified for the implementation of this desired change. Such strategy is generally determined by client and helper working together.

Implementational support and strategy identification clearly lend themselves to the incorporation of a wide variety of behavioral techniques and approaches. Brief general strategy identification guidelines, which may be of use in providing implementational support, are shown below:

Strategy Identification Guidelines

1. Keep It Simple.
2. Take Small Steps.
3. Make It Replicable.
4. Focus on Positive Action.
5. Be Independent.
6. Make Sure It Is Reasonable.

---

1. *Keep it simple.*   Encourage clients to develop a strategy that's simple and easy to implement. Simple strategies are easier to remember and more likely to be maintained. For example, a client may identify a goal and desired behavior change as "assertively expressing my feelings."

2. *Take small steps.*   Encourage clients to break their goals into small sequential steps. For example, the goal of being assertive can be broken down into (1) becoming aware of how they feel, (2) stating to themselves how they feel, (3) stating aloud how they feel, (4) making an owning statement of how they feel and what they need, and (5) making a complete assertive statement including recognition of what the other person needs, feels, and wants. Encourage clients to focus on step 1 first.

3. *Make it replicable.*   Whatever the strategy is, it must be replicable. The client has had years of practice doing the undesired behavior and, consequently, it's easier to default to that behavior (e.g., being unassertive). The plan for developing the new desired behavior (being assertive) must be simple enough and a small enough step that it can be repeated over and over again in many different and often challenging situations. The idea is to develop a new *pattern* of behavior (the desired behavior). That behavior, therefore, must be replicable.

4. *Focus on positive action.*   Often we have a tendency to focus on what we *don't* want to do: "Don't panic," "Don't lose your temper." It's much more useful and helpful, however, to focus on what we *do* want to do: "Concentrate on breathing," "Count to ten." If we want to stop doing one thing, we need to have something to replace it with; otherwise we default to our familiar behavior pattern.

5. *Be independent.*   Whatever strategy your client develops, it needs to be one that he or she can do in any setting, in any situation, and independently. Sometimes we'd like to put the responsibility for behavior change onto someone else. For example, "Whenever you see me being unassertive, remind me to say something." It is permissible to enlist support, but the ultimate responsibility rests with the client, who must be able to do the behavior change without the help of someone else. Otherwise, the change is dependent on others.

6. *Make sure it is reasonable.*   Reasonable means reasonable *to the client.* The change must be something clients want; otherwise it will not happen. If they are having trouble implementing the change, then

perhaps (1) they have conflicting goals—that is, they want both to be assertive and to avoid the rejection they fear, or (2) they do not want to make the change—it's something somebody else, such as a partner or parent, *wants* them to want. At that point, it's time to return to Stage 2 and reexamine the issues.

One must possess important clinical skills that help assist individuals in finding the impediments that block their ability to undergo some changes and function at a more effective higher level. (adapted from Doyle, 1992)

Effective implementation of a strategy may be influenced by 3 factors: sequencing of strategies; rationale for a strategy; and instructions about a strategy. (Hackney & Cormier, 1988, p. 121)

Strategy Implementation

Once the strategy for the desired change is identified, using the guidelines above, the next step is to implement it. The client now needs to begin *doing* it. Apply the strategy as it has been developed. To bring about the desired goal and change, the client must ultimately commit to *doing* the new behavior.

Implementation should incorporate successive steps allowing the client to move toward the goal and desired change gradually. For example, clients with a goal of "assertively expressing their feelings" may want to begin by simply stating their feelings to themselves. They might then state them out loud to themselves, and then out loud to someone they know is supportive. Finally, they might build up to stating their feelings in more challenging situations, such as in a group of people they don't know, or with individuals they expect to be unsupportive.

It may be helpful for clients to keep a journal or log of their strategy implementation. This can be useful both to monitor and to encourage consistent implementation—e.g., commitment to assertively expressing feelings at least once a day—and for assessment of progress and success.

Strategy Assessment

Evaluating the success of a strategy is important. If the implementation of a strategy is evaluated as unsuccessful, the steps identified under strategy identification should be reassessed. It may be that one or more steps of the general strategy identification guideline need to be reworked. Pay particular attention to step 6: make sure it's reasonable. If clients have difficulty implementing a strategy, perhaps they don't perceive the strategy and/or goal as reasonable. If not, they need to go back and reassess their identified goal.

Potential problem areas that can interfere with implementational support include these:

1. Lack of direction: Without identified direction, no strategy can be identified.
2. Lack of strategy identification: Without a decision regarding a strategy for implementation, no plan can be formed.
3. Lack of action: Lack of follow-up by the client on the plan of action.

HELPING RELATIONSHIP GOALS

The goal for the helping relationship in Stage 1 was rapport and trust, for Stage 2 it was processing and understanding, and for Stage 3 it is to provide an environment for directionality and change. The process goal is directionality. The outcome goal is change (see Part V Table).

Directionality

Directionality as a process goal of the relationship refers to the emergence of direction within the helping relationship. The helper facilitates the emergence of that direction. Directionality emerges as a result of (1) the foundational three R's, and (2) the effective implementation of the Stage 1, 2, and 3 counselor process skills to facilitate the Stage 1, 2, and 3 client process goals (see Part V Table).

Directionality through the Three R's

The beliefs and attitudes regarding the clients' rights, self-respect, and self-responsibility continue to form the foundation for both effective processing and direction (see Part III).

Directionality through Implementation of Process Skills

Directionality is the result of the cumulative helper communication and related helper process skills. The effective application of these skills helps facilitate client participation in client growth goals of (1) direction, (2) strategy and implementation, and (3) change.

Change

The ultimate outcome goal is change, representing growth toward the client's self-understanding, new perspective, and corresponding desired behavioral change. Although sessions may terminate prior to the full manifestation of this change, theoretically, the change in the client will be observable within the helping relationship.

SUMMARY

Stage 3 focuses on direction and change and includes the period of time from the development of new perspectives and insight through actual behavioral change.

From the clients' perspective, Stage 3 includes determination of direction, implementation of strategy, and behavioral change. Helper skills in Stage 3 incorporate the fifth step of the communication model and the following corresponding facilitative skills:

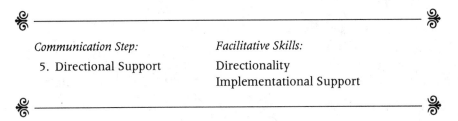

Communication Step:	Facilitative Skills:
5. Directional Support	Directionality Implementational Support

Relationship goals in Stage 3 include directionality and change.

Epilogue:
The Framework
in Review

P urpose and goals must be clear in order to recognize where you are going, and when and whether you get there. The purpose of communication in a helping relationship is to help the client grow toward improved psychological health through increased self-understanding. More specifically, the goal is growth (process goal) toward improved psychological health as evidenced by, in this framework, a positive self-concept (outcome goal). The purpose of the helping relationship is to facilitate achievement of these goals.

The helping relationship needs to have a structural framework within which the helper can operate. In addition to clarity of purpose, it needs a philosophy of growth for self and others, communication skills, and facilitative skills.

The philosophy of growth, presented here as an outcome goal for the client, helper, and helping relationship, suggests a foundation of underlying beliefs regarding the *right* to be oneself, *respect* for individuals' unique capabilities and differences, and *responsibility* for one's actions and choices. This foundation of beliefs is translated into behaviors of congruence, competence, and internalized control, respectively. These behaviors reflect an affective experience of self-acceptance, self-esteem, and self-actualization.

The philosophy also suggests that these beliefs for oneself are expressed as attitudes concerning rights, respect, and appropriate responsibility-taking when applied to clients. Behaviorally, the helper with these attitudes is genuine, gives positive regard, and is able to keep clear boundaries through focusing and appropriate responsibility-taking. The

helper, in so doing, becomes effective by role modeling, by being a catalyst, and by being an effective facilitator. The stage structure for the
helping process includes pre-stage attitudes, communication skills, and
facilitative skills emerging in sequential stages.

PRE-STAGE ATTITUDES

Attitudes speak louder than words. Words cannot substitute for attitude. Attitude, therefore, is the most powerful message we communicate. Verbal skills are useless without effective attitude. An attitude reflecting belief in the clients' *right* to be who they are, showing *respect* for
their unique capabilities and differences, and encouraging appropriate
responsibility-taking for self and client are therefore suggested as fundamental pre-stage attitudes.

COMMUNICATION SKILLS

Communication skills are the core of the helping process. Since patterns of communication have already been developed, the question is
how they can be used consistently and effectively in facilitating the personal growth of the client.

FACILITATIVE SKILLS

Facilitative skills are the effective use of communication skills for the
purpose of promoting growth. A structure of stages and skills is a systematic, operational organization of facilitative skills.

THE STAGE STRUCTURE

The three stages of the helping process framework generally correspond to the three basic human dimensions of affective experience

(feeling and being), behavior (action and doing), and cognition (thinking and choosing).

In Stage 1 (the presenting problem), the focus is on validating and affirming the clients' feelings and accepting who they are and where they are. In Stage 2 (the underlying issues), the focus shifts to encouraging thinking and choosing. Clients are assisted in the cognitive processes of analyzing, examining underlying issues, and developing new perspectives. With new perspective and insight comes a decision, a choice in determining direction. Finally, in Stage 3 (direction and change), the focus shifts again, to implementation of what clients have decided to do. In other words, Stage 1 highlights "where the client is" (being); Stage 2, "where the client wants to go" (choosing), and Stage 3, "the going" (doing).

The stages also parallel the three R's—rights, respect, and responsibility—and their respective behavioral components of congruence, competence, and internalized control. Stage 1 and its focus on acceptance of clients being and affective experience, reflects the belief and attitude of clients' right to be who they are. The helper's attitude reflects acknowledgment of the individuals' right to be who they are with their own unique perspective, and is an invitation for clients to be congruent. Stage 2, and its focus on choosing and thinking, reflects the belief and attitude of respect for clients' ability to figure out who they are and what they need. It reflects a belief in and promotion of clients' competence. Finally, Stage 3, with its focus on doing and behaving, reflects the belief and attitude that we are responsible for ourselves, our actions, and our choices. With the right to be who we are comes the responsibility of who we are, and what we need to feel good about ourselves and how we're living our lives. Appropriate self-responsibility reflects internalized control. Together these stages symbolize the overall outcome goal of client growth: psychological health, as evidenced by a positive self-concept—the ultimate purpose of the helping relationship.

References

Adler, A. (1963). *The practice and theory of individual psychology* (P. Radin, Trans.). Paterson, NJ: Littlefield, Adams.

American Counseling Association. (1988). *Ethical standards* (rev. ed.). Alexandria, VA: Author.

American Psychological Association. (1981). *Ethical principles of psychologists* (rev. ed.). Washington, DC: Author.

Arkoff, A. (1988). The meaning of personal growth. In A. Arkoff (Ed.), *Psychology and personal growth* (3rd ed., pp. 331–338). Boston: Allyn & Bacon.

Beck, A. (1976). *Cognitive therapy and the emotional disorders.* New York: International Universities Press.

Beck, A., & Weishaar, M. E. (1989). Cognitive therapy. In R. J. Corsini & D. Wedding (Eds.), *Current psychotherapies* (pp. 285–320). Itasca, IL: F. E. Peacock.

Belkin, G. S. (1988). *Introduction to counseling* (3rd ed.). Dubuque, IA: William C. Brown.

Brammer, L. M. (1988). *The helping relationship: Process and skills* (4th ed.). Englewood Cliffs, NJ: Prentice-Hall.

Brammer, L. M., Shostrom, E. L., & Abrego, P. J. (1989). *Therapeutic psychology: Fundamentals of counseling and psychotherapy* (5th ed.). Englewood Cliffs, NJ: Prentice-Hall.

Bruch, H. (1981). Teaching and learning psychotherapy. *Canadian Journal of Psychiatry, 26,* 86–92.

Bugental, J. F. T. (1987). *The art of the psychotherapist.* New York: Norton.

Burke, J. F. (1989). *Contemporary approaches to psychotherapy and counseling.* Pacific Grove, CA: Brooks/Cole.

Burr, W. (1990). Beyond I-statements in family communication. *Family Relations, 39,* 266–273.

Carkhuff, R. R. (1969). *Helping and human relations: Vol. 1. Selection and training.* New York: Holt, Rinehart & Winston.

Carmin, C. N., & Dowd, T. E. (1988). Paradigms in cognitive psychotherapy. In W. Dryden & P. Trower (Eds.), *Developments in cognitive psychotherapy* (pp. 1–20). Beverly Hills, CA: Sage.

Carter, M. J., Van Ardel, G. E., & Robb, G. M. (1985). *Therapeutic recreation: A practical approach.* Prospect Heights, IL: Waveland Press.

Chebat, J., & Picard, J. (1988). Receivers' self-acceptance and the effectiveness of two-sided messages. *Journal of Social Psychology, 128,* 353–362.

Conway, T. L., Vickers, R. R., & French, J. R. P. (1992). An application of person–environment fit theory: Perceived versus desired control (part of a symposium on the heritage of Kurt Lewin). *Journal of Social Issues, 48,* 95–107.

Corey, G. (1991). *Theory and practice of counseling and psychotherapy* (4th ed.). Pacific Grove, CA: Brooks/Cole.

Corey, G., Corey, M. S., & Callahan, P. (1993). *Issues and ethics in the helping professions* (4th ed.). Pacific Grove, CA: Brooks/Cole.

Cormier, W. H., & Cormier, L. S. (1991). *Interviewing strategies for helpers: Fundamental skills and cognitive behavioral interventions* (3rd ed.). Pacific Grove, CA: Brooks/Cole.

Corsini, R. J., & Wedding, D. (1989). *Current psychotherapies* (4th ed.). Itasca, IL: F. E. Peacock.

Cottone, R. (1992). *Theories and paradigms of counseling and psychotherapy.* Boston: Allyn & Bacon.

Covey, G. (1990). *The seven habits of highly effective people.* New York: Simon & Schuster.

Craig, G. (1989). *Human development* (5th ed.). Englewood Cliffs, NJ: Prentice-Hall.

Cushman, P. (1990). Why the self is empty. *American Psychologist, 45*(5), 599–611.

Daniels, M. (1988). The myth of self-actualization. *Journal of Humanistic Psychology, 28*(1), 7–38.

Davison, G. C., & Neale, J. M. (1986). *Abnormal psychology* (4th ed.). New York: Wiley.

Deffenbacher, J. L. (1985). A cognitive-behavioral response and a modest proposal. *Counseling Psychologist, 13,* 261–269.

Doyle, E. R. (1992). *Essential skills and strategies in the helping process.* Pacific Grove, CA: Brooks/Cole.

Egan, G. (1990). *The skilled helper: A systematic approach to effective helping* (4th ed.). Pacific Grove, CA: Brooks/Cole.

Egan, G. (1994). *The skilled helper: A systematic approach to effective helping* (5th ed.). Pacific Grove, CA: Brooks/Cole.

Ellis, A. (1989). Rational-emotive therapy. In R. J. Corsini & D. Wedding (Eds.), *Current psychotherapies* (pp. 197–238). Itasca, IL: F. E. Peacock.

Ellis, A., & Harper, A. (1975). *A new guide to rational living.* Hollywood, CA: Wilshire Book Co.

Evans, R. D., Hearn, T. M., Uhlman, R. M., & Ivey, E. A. (1989). *Essential interviewing: A programmed approach to effective communication* (3rd ed.). Pacific Grove, CA: Brooks/Cole.

Fadiman, J., & Fraser, R. (1984). *Personality and personal growth* (2nd ed.). New York: Harper & Row.

Fisch, R., Weakland, J., & Segal, L. (1985). *The tactics of change: Doing therapy briefly.* San Francisco: Jossey-Bass.

Frank, J. (1971). Therapeutic factors in psychotherapy. *American Journal of Psychotherapy, 25,* 350–361.

Frankl V. (1985). *The unheard cry for meaning: Psychotherapy and humanism.* New York: Simon & Schuster.

Fromm, E. (1982). *Escape from freedom.* New York: Avon.

Gazda, M. G., Asbury, S. F., Balzer, J. F., Childres, C. W., & Walters, P. R. (1984). *Human relations development: A manual for educators* (3rd ed.). Boston: Allyn & Bacon.

George, R. L., & Cristiani, T. S. (1990). *Counseling: Theory and practice.* Englewood Cliffs, NJ: Prentice-Hall.

Gladding, S. T. (1988). *Counseling: A comprehensive profession.* Columbus, OH: Merrill.

Gladding, S. T. (1992). *Counseling: A comprehensive profession* (2nd ed.). New York: Macmillan.

Glasser, W. (1965). *Reality therapy.* New York: HarperCollins.

Goldin, E., & Doyle, R. (1991). Counselor predicate usage and communication proficiency on ratings of counselor empathic understanding. *Counselor Education and Supervision, 30,* 212–224.

Gorden, L. R. (1992). *Basic interviewing skills.* Itasca, IL: F. E. Peacock.

Gordon, T. (1970). *Parent effectiveness training: The tested new way to raise responsible children.* New York: Peter H. Wyden.

Hackney, H., & Cormier, L. S. (1988). *Counseling strategies and interventions* (3rd ed.). Englewood Cliffs, NJ: Prentice-Hall.

Hall, C. H., & Lindzey, G. (1985). *Introduction to theories of personality.* New York: Wiley.

Hancock, E. (1981). Women's development in adult life. *Dissertation Abstracts International 42*(6), 2504.

Heylighen, F. (1992). A cognitive-systematic reconstruction of Maslow's theory of self-actualization. *Behavioral Science, 37,* 39–58.

Hines, M. H. (1988). Whose problem is it? *Journal of Counseling and Development, 67,* 106.

Horney, K. (1945). *Our inner conflicts.* New York: Norton.

Humes, C. W. (1987). *Contemporary counseling.* Muncie, IN: Accelerated Development.

Hurley, J. R. (1989). Self-acceptance and other-acceptance scales for small groups. *Genetic, Social and General Psychology Monographs, 115*(4), 483–503.

Hutchins, E. D., & Cole, G. C. (1992). *Helping relationships and strategies* (2nd ed.). Pacific Grove, CA: Brooks/Cole.

Ivey A., Ivey, M., & Simek-Morgan, L. (1993). *Counseling and psychotherapy: A multicultural perspective.* Boston: Allyn & Bacon.

Ivey, E. A. (1991). *Development strategies for helpers: Individual, family, and network interventions.* Pacific Grove, CA: Brooks/Cole.

Ivey, E. A. (1994). *Intentional interviewing and counseling: Facilitating client development* (3rd ed.). Pacific Grove, CA: Brooks/Cole.

Ivey, E. A., & Simek-Downing, L. (1980). *Counseling and psychotherapy: Skills, theories, and practice.* Englewood Cliffs, NJ: Prentice-Hall.

John-Roger, J., & McWilliams, P. (1991). *Life 101.* Los Angeles: Prelude Press.

Johnson, D. (1993). *Reaching out.* Needham Heights, MA: Allyn & Bacon.

Joubert, C. E. (1990). Relationship among self-esteem, psychological reactance, and other personality variables. *Psychological Reports, 66,* 1147–1151.

Kazdin, A. (1977). *The token economy: A review and evaluation.* New York: Plenum.

Kottler, J. A., & Brown, R. W. (1992). *Introduction to therapeutic counseling* (2nd ed.). Pacific Grove, CA: Brooks/Cole.

Leviton, R. (1992, September/October). Reconcilable differences. *Yoga Journal,* pp. 48–104.

Løvlie, A. (1982). *The self of the psychotherapist: Movement and stagnation in psychotherapy.* Oslo—Bergen—Tromsø: Universitetsforlaget.

Maslow, A. (1954). *Motivation and personality.* New York: Harper & Row.

May, R. (1989). *The art of counseling* (rev. ed.). New York: Gardner Press.

McGrew, K. S., & Bruininks, R. H. (1990). Defining adaptive and maladaptive behavior within a model of personal competence. *School Psychology Review, 19*(1), 53–73.

Meichenbaum, D. (1991). Evolution of cognitive behavior therapy. In J. Zeig (Ed.), *The evolution of psychotherapy, II.* New York: Brunner/Mazel.

Mikulas, W. L. (1978). *Behavior modification.* New York: Harper & Row.

Millman, D. (1984). *Way of the peaceful warrior.* Tiburon, CA: H. J. Kramer.

Mittleman, W. (1991). Maslow's study of self-actualization: A reinterpretation. *Journal of Humanistic Psychology, 31,* 114–135.

Moursund, J. (1990). *The process of counseling and therapy* (2nd ed.). Englewood Cliffs, NJ: Prentice-Hall.

National Board for Certified Counselors. (1989). *National Board for Certified Counselors code of ethics.* Alexandria, VA: Author.

National Board for Certified Counselors. (n.d.). *Your consumer guide to counseling services: Client rights and responsibilities.* Alexandria, VA: Author.

Nelson-Jones, R. (1993). *Lifeskills helping.* Pacific Grove, CA: Brooks/Cole.

O'Connor, G. (1993). *The aikido student handbook.* Berkeley, CA: Frog, Ltd./North Atlantic Books.

Okun, F. B. (1992). *Effective helping: Interviewing and counseling techniques* (4th ed.). Pacific Grove, CA: Brooks/Cole.

Osipow, S. H., & Betz, N. E. (1991). Career counseling research. In C. E. Watkins, Jr., & L. J. Schneider (Eds.), *Research in counseling* (pp. 205–233). Hillsdale, NJ: Erlbaum.

Patterson, C. H. (1986). *Theories of counseling and psychotherapy* (4th ed.). New York: Harper & Row.

Patterson, L. D., & Eisenberg, S. (1983). *The counseling process* (3rd ed.). Boston: Houghton Mifflin.

Patterson, L. M. (1983). *Nonverbal behavior: A functional perspective*. New York: Springer-Verlag.

Perls, F. (1969). *Gestalt therapy verbatim*. Menlo Park, CA: Real People Press.

Phares, E. J. (1992). *Clinical psychology: Concepts, methods, and profession* (4th ed.). Pacific Grove, CA: Brooks/Cole.

Raskin, N., & Rogers, C. R. (1989). Person-centered therapy. In R. Corsini & D. Wedding (Eds.), *Current psychotherapies* (4th ed.). Itasca, IL: Peacock.

Rogers, C. R. (1958). The characteristics of a helping relationship. *Personnel and Guidance Journal, 37,* 6–16.

Rogers, C. R. (1961). *On becoming a person: A therapist's view of psychotherapy*. Boston: Houghton Mifflin.

Rogers, C. R. (1967). The necessary and sufficient conditions of the therapeutic personality change. *Journal of Counseling, 21,* 95–103.

Rogers, C. R. (1980). *A way of being*. Boston: Houghton Mifflin.

Rogers, C. R., & Wallen, J. (1946). *Counseling with returned servicemen*. New York: McGraw-Hill.

Ryff, C. D. (1989). Happiness is everything, or is it? Explorations on the meaning of psychological well-being. *Journal of Personality and Social Psychology, 57*(6), 1069–1081.

Sanders, J., Jones, E., & Sanders, R. C. (1987). Human relations laboratory groups for enhancing personal growth and self-discovery among graduate students. *College Student Journal, 21*(3), 249–253.

Satir, V. (1988). *The new peoplemaking*. Mountain View, CA: Science and Behavior Books.

Schmidt, L. D. (1980). Why has the professional practice of psychological counseling developed in the United States? In J. M. Whitney & B. R. Fretz (Eds.), *The present and future of counseling psychology* (pp. 29–33). Pacific Grove, CA: Brooks/Cole.

Schultz, D. (1990). *Theories of personality* (4th ed.). Pacific Grove, CA: Brooks/Cole.

Shostrom, E. (1974). *Personal orientation inventory*. San Diego, CA: Educational and Industrial Testing Service.

Strong, S. R., & Claiborn, C. D. (1982). *Change through interactions: Social psychological processes of counseling and psychotherapy*. New York: Wiley.

Teyber, E. (1992). *Interpersonal process in psychotherapy*. Pacific Grove, CA: Brooks/Cole.

Thompson, G., & Stroud, M. (1984). *Verbal judo*. Albuquerque, NM: Communication Strategies.

Van Houten, R., Axelrod, S., Bailey, J. S., & Favell, J. (1988). The right to effective behavioral treatment. *Behavior Analyst, 11*(2), 111–114.

Verderber, R. F., & Verderber, K. S. (1989). *Inter-act: Using interpersonal communication skills* (5th ed.). Belmont, CA: Wadsworth.

Waite, B. T., Gansneder, B., & Rotella, R. J. (1990). A sport-specific measure of self-acceptance. *Journal of Sport and Exercise Psychology, 12*(3), 264–279.

Warton, P. M., & Goodnow, J. J. (1991). The nature of responsibility: Children's understanding of "your job." *Child Development, 62*(1), 156–165.

Webster's New Collegiate Dictionary (8th ed.). (1981). Springfield, MA: G. & C. Merriam.

Wells, R. A. (1982). *Planned short-term treatment.* New York: Free Press.

Williamson, M. (1992). *A return to love.* New York: HarperCollins.

Wolpe, J. (1958). *Psychotherapy by reciprocal inhibition.* Stanford, CA: Stanford University Press.

Woody, R. H., Hansen, C. J., & Rossberg, H. R. (1989). *Counseling psychology: Strategies and services.* Pacific Grove, CA: Brooks/Cole.

Wright, B. A. (1987). Human dignity and professional self-monitoring: Professionalism and futurism [Special issue]. *Journal of Applied Rehabilitation Counseling, 18*(4), 12–14.

Yahne, C., & Long, V. (1988). The use of support groups to raise self-esteem for women clients. *Journal of American College Health, 37,* 79–84.

Yalom, I. D. (1980). *Existential psychotherapy.* New York: Basic Books.

Yontef, G. M., & Simkin, J. S. (1989). Gestalt therapy. In R. Corsini & D. Wedding (Eds.), *Current psychotherapies* (4th ed., pp. 323–361). Itasca, IL: Peacock.

Index

To the owner of this book:

I hope that you have enjoyed *Communication Skills in Helping Relationships* as much as I have enjoyed writing it. I'd like to know as much about your experiences with the book as you care to offer. Only through your comments and the comments of others can I learn how to make a better book for future readers.

School: _____

Your Instructor's name: _____

1. For what course was this book assigned? _____

_____

_____

2. What did you like most about *Communication Skills in Helping Relationships?*

_____

_____

3. What did you like least about the book? _____

_____

_____

4. Were all of the chapters of the book assigned for you to read? _____

   If not, which ones weren't? _____

_____

5. In the space below, or in a separate letter, please let me know what other comments about the book you'd like to make. (For example, were any chapters or concepts particularly difficult?) I'd be delighted to hear from you!

_____

_____

_____

_____

Optional:

Your name: _____ Date: _____

May Brooks/Cole quote you either in promotion for *Communication Skills in Helping Relationships* or in future publishing ventures?

Yes: _____ No: _____

Sincerely,

Vonda Long

Brooks/Cole Publishing is dedicated to publishing quality publications for education in the human services fields. If you are interested in learning more about our publications, please fill in your name and address and request our latest catalogue, using this prepaid mailer.

Name: _____

Street Address: _____

City, State, and Zip: _____

FOLD HERE

FOLD HERE